Basic Statistics in the Human Services

BASIC STATISTICS
IN THE
HUMAN SERVICES
an Applied Approach

by
Ann E. MacEachron, M.S.W., Ph.D.

Associate Professor, Florence Heller Graduate School
Brandeis University

Director of the Sociology Research Department
Eunice Kennedy Shriver Center for Research on Mental Retardation

Director of the Program Research Unit
New York State Office of Mental Retardation and Developmental Disabilities

University Park Press
Baltimore

UNIVERSITY PARK PRESS
International Publishers in Science, Medicine, and Education
300 North Charles Street
Baltimore, Maryland 21201

Copyright© 1982 by University Park Press

Typeset by University Park Press Typesetting Division

Manufactured in the United States of America by
The Maple Press Company

Library of Congress Cataloging in Publication Data

MacEachron, Ann E.
Basic statistics in the human services.

Includes index.
1. Social sciences—Statistical methods. 2. Social service—
Statistical methods. 3. Statistics.   I. Title.
HA29.M174   519.5   81–19789
ISBN 0–8381–1728–0   AACR2

Contents

APPENDIX: STATISTICAL TABLES

Preface

The purpose of this book is to provide an introduction to the basics of statistics for students who are being trained in the various areas of human services, at the undergraduate or master's level. A second purpose is to design a text for easy reference long after a student has taken the course. Few human services professionals care to practice the skills of research and statistics on a daily basis, but in the course of their work they may need to understand a certain statistic.

Three approaches are used to accomplish these goals. First, decision-making charts to guide the selection of appropriate statistics are presented in almost every chapter. These charts summarize the logic underlying the choice between statistics and also reference the section where each choice is discussed. Second, emphasis is placed on the conceptual understanding and interpretation of statistics rather than on their mathematical foundations. Third, emphasis is placed on computer analysis of statistics rather than on hand calculation. Because almost all analysis is now done by computers, and because the process of de-mystifying both statistics and computers is helpful in learning, most statistics discussed in this book are also defined in terms of how to do them by computers. In this regard, where appropriate, chapters of the *SPSS Primer* are referred to and their programs are used for demonstration purposes.

The book begins with three chapters on basic knowledge for beginning the study of statistics. Chapter 1 provides a brief introduction to the role of statistics in professional practice and to basic definitions used in statistics. Chapter 2 reviews the arithmetic needed to calculate statistics, and Chapter 3 introduces some elementary statistical notation. Part II of the book focuses on descriptive statistics and ends with suggestions on how to write a research report (Chapter 12) and problems for practicing the use of descriptive statistics (Chapter 13). To facilitate the learning of and choice among descriptive statistics, examples from one study—a hypothetical detention home—are used throughout this part of the book. It should be noted as well that the *SPSS Primer* provides two data-based examples that could also be used for class purposes. Part III focuses on inferential statistics. Chapter 14 presents the basic concepts necessary to understand the rationale of hypothesis testing. Chapter 15 describes 13 hypothesis tests for univariate and bivariate analyses commonly done in the human services and social sciences.

I am indebted to my previous statistics teachers as well as my students. My previous teachers led me to understand both the logic and the meaning of statistical analysis. My students demanded a sense of applied purpose for learning statistics and a clarity of presentation. Any faults of this book, therefore, are solely my responsibility.

I am also most grateful to Pat List, Rita McKeown, and Ina Moses for preparing the manuscript so carefully and patiently.

To Jim and Jimmy

Basic Statistics in the Human Services

Chapter 1

Introduction

This book is written from the perspective that human services professionals need to know at least a little about statistics. Statistics are an integral part of the growing research literature that evaluates client outcomes in different treatment interventions. If professional knowledge is to be improved, an understanding of this literature is necessary. Statistics are part of the increasing demand for accountability of service effectiveness and cost efficiency. Statistics are part of any agency's reporting system to apply for continued program funding. Statistics, in short, play an important role in the human services.

Yet few human services professionals care to deal with statistics, especially the mathematical derivations and definitions of statistics. This book uses no more math or statistical notation than that presented in Chapters 2 and 3. The remaining chapters focus on presenting and explaining statistics in English. The goal of this approach is to provide a conceptual understanding of statistics and a basic facility in appropriately applying and interpreting statistics on an applied basis.

Statistics may perhaps be best understood as a way of dealing with information, observations, or data that have an empirical base. In any empirical study, information is collected. This information is patterned, observable, and variable. Patterned information means that events may be characterized as following systematic laws or repeatable patterns that are potentially observable. Observable information means that independent observers or researchers can see and measure the same events. Variable information means that the event being observed is capable of changing, varying, or differing. It is this last characteristic that has come to label the information analyzed in statistics. A *variable* is a characteristic of a person, object, or thing that is observable and that varies. It is presumed to reflect the event of interest—usually a concept such as leadership, poverty, or health—and thus a variable or set of variables may also be called the operationalization of the concept. *Operationalization* is the set of instructions used to translate an abstract concept to an observable variable(s). These instructions tell you how to observe and measure the event and its changing values. Poverty, for example, may be opera-

1

tionalized by the instructions to observe and measure a person's income in dollars. The result of operationalizing a concept is a variable measured by a scale with different categories.

Measurement is the assignment of numerical values to observed phenomena. These values or categories must be mutually exclusive and exhaustive; that is, each observed event may only be classified into one category, and each observed event must have a category into which it may be properly classified. *Nominal-level* measurement scales refer to numerical values that represent only the classification of events: black/white, male/female, on-welfare/off-welfare, or the numbers on football uniforms. These numerical values, therefore, serve only as labels and permit no arithmetic operations. *Ordinal-level* measurement scales refer to numerical values that represent both the classification and rank-ordering of events: rank in high school class, grades on exams ranging from A to F, or the extent of belief in an attitude where a score of 1 indicates strong disbelief and a score of 5 indicates strong belief. Although ordinal values represent ranked values, the ranks may be of different magnitudes or sizes. Thus, only operations which do not alter the order of ranks are permissible. *Interval-level* measurement scales refer to numerical values that represent classifications, rank ordering, and equality of the size of the intervals or ranks: number of gallons of gasoline, IQ scores, minutes or hours, or amount of dollars received on a welfare check. The equality of intervals permits not only a comparison of scores between subjects as for ordinal scales, but also permits an analysis of how many intervals or units more (or less) one subject is than another subject. Not only does one family have more children than another family, for example, but that family has three more children.

These distinctions between scales of measurement for variables are important for choosing which statistic to use in analyzing data. Before demonstrating this, however, the basics of math and statistical notation underlying the presentation of such statistics is introduced. The book is then divided into two additional parts, one part for descriptive statistics and the other part for inferential statistics. *Descriptive statistics* are methods that organize and summarize only the observed collected information. *Inferential statistics* are methods that may be used to generalize from the observed information to a larger group or population that has not been directly observed.

PART I

BASICS

Chapter 2

Review of
Basic Mathematics

This chapter reviews basic mathematics. This is the only mathematics required to use the statistics presented in this book.

2.1 ADDITION

When all numbers to be added are positive, add them without regard to sign. Either attach a positive sign (+) to the result or leave the result without a sign, since no sign always indicates a positive number.

Examples	$(12) + (6)$	$(13) + (0)$	$(5) + (19)$
Rewritten	12	13	5
	6	0	19
Answers	18	13	24

When all numbers to be added are negative, add them without regard to sign. Attach a negative sign (−) to the result.

Examples	$(-12) + (-6)$	$(-13) + (-0)$	$(-5) + (-19)$
Rewritten	12	13	5
	6	0	19
Answers	−18	−13	−24

When the numbers to be added have different signs, first add all the numbers with positive signs together and then add all the numbers with negative signs together. Disregard the positive or negative signs of each sum, and subtract the larger sum from the smaller sum to get the result. If the larger sum is positive, then the result is positive and no sign is needed. If the larger sum is negative, then the result is negative and a negative sign must be attached.

Examples	$(12) + (-6) + (-3)$	$(-12) + (6) + (3)$	$(12) + (-6) + (3)$
Rewritten	$12 + (-9)$	$(-12) + 9$	$15 + (-6)$
Answers	3	−3	9

2.2 SUBTRACTION

To subtract a number from another number, change the original sign of the number to be subtracted (always the number to the right of the minus sign) to its opposite sign. That is, change an original positive sign to a negative sign and an original negative sign to a positive sign. Then follow addition rules to get the result or remainder.

Examples	$(12)-(6)$	$(13)-(0)$	$(5)-(19)$
Rewritten	$\begin{array}{r}12\\-6\\\hline\end{array}$	$\begin{array}{r}13\\-0\\\hline\end{array}$	$\begin{array}{r}5\\-19\\\hline\end{array}$
Answers	6	13	-14

Examples	$(5)-(-19)$	$(-5)-(19)$	$(-5)-(-19)$
Rewritten	$\begin{array}{r}5\\+19\\\hline\end{array}$	$\begin{array}{r}-5\\-19\\\hline\end{array}$	$\begin{array}{r}-5\\+19\\\hline\end{array}$
Answers	24	-24	14

2.3 MULTIPLICATION

To multiply two whole numbers (integers), multiply the numbers without regard to sign. The result, or product, is positive if both numbers are positive or both numbers are negative. The result or product is negative if one number is positive and the other number is negative.

Examples	$(12)\times(6)$	$(5)\times(19)$	$(5)\times(-19)$	$(-5)\times(19)$	$(-5)\times(-19)$
Rewritten	$\begin{array}{r}12\\6\\\hline\end{array}$	$\begin{array}{r}5\\19\\\hline\end{array}$	$\begin{array}{r}5\\-19\\\hline\end{array}$	$\begin{array}{r}-5\\19\\\hline\end{array}$	$\begin{array}{r}-5\\-19\\\hline\end{array}$
Answers	72	95	-95	-95	95

Any number multiplied by zero gives a product of zero.

Examples	$(12)\times(0)$	$(5)\times(0)$	$(0)\times(3)$
Rewritten	$\begin{array}{r}12\\0\\\hline\end{array}$	$\begin{array}{r}5\\0\\\hline\end{array}$	$\begin{array}{r}0\\3\\\hline\end{array}$
Answers	0	0	0

To multiply numbers with decimals, first multiply the numbers without regard to their decimal points. To determine where to place the decimal point in the resulting product, count and add the total number of digits to the right of the decimal point for each original number. This total indicates the number of spaces that should be on the right of the decimal point in the product.

Examples	$.03\times.009$	$.72\times.0001$	2.1×5.3

Rewritten	3	72	21
	$\times 9$	$\times 1$	$\times 53$
Multiply	27	72	1113
Count from deci- mal point	$2+3=5$	$2+4=6$	$1+1=2$
Answers	$.00027$	$.000072$	11.13

To multiply a fraction by a whole number, multiply the fraction's numerator (the top or first number in a fraction) by that whole number.

Examples	$(\tfrac{3}{4})\times(8)$	$(\tfrac{1}{2})\times(-10)$	$(\tfrac{1}{3})\times(0)$
Rewritten	$\dfrac{(3\times 8)}{4}$	$\dfrac{(1\times -10)}{2}$	$\dfrac{(1\times 0)}{3}$
Answers	6	-5	0

To multiply two fractions, first multiply the two numerators together and then multiply the two denominators together.

Examples	$(\tfrac{3}{4})\times(\tfrac{1}{2})$	$\dfrac{-5}{8}\times\dfrac{2}{5}$	$\dfrac{-2}{7}\times\dfrac{-3}{8}$
Rewritten	$\dfrac{(3\times 1)}{(4\times 2)}$	$\dfrac{(-5\times 2)}{(8\times 5)}$	$\dfrac{(-2\times -3)}{(7\times 8)}$
Answers	$\dfrac{3}{8}$	$\dfrac{-10}{40}$	$\dfrac{6}{56}$

2.4 DIVISION

To divide two whole numbers, divide the numbers without regard to sign. The result, or quotient, is positive if both numbers were positive or both numbers were negative. The result is negative if one number was positive and the other number was negative. There are three ways of expressing division:

1) Divisor $\overline{)\text{dividend}}^{\text{quotient}}$
2) Dividend $\div$ divisor = quotient
3) Dividend / divisor = quotient

Examples	$2\overline{)16}$	$16/2$	$-16/2$	$2/-16$	$16\div 2$
Answers	8	8	-8	$-.125$	8

If you divide zero by any other number, the result or quotient is always zero.

Examples	$2\overline{)\,0}$	$0/2$	$0 \div 100$
Answers	0	0	0

If you divide any number by zero, the result is undefined or meaningless.

Examples	$0\overline{)\,2}$	$2/0$	$100/0$	$0 \div 0$
Answers	undefined	undefined	undefined	undefined

To divide decimals, do the following steps: 1) move the decimal point of the divisor enough spaces to the right to make the divisor a whole number, 2) move the decimal point of the dividend the same number of spaces to the right, 3) put the decimal point of the quotient above the new decimal point in the dividend, and 4) divide the numbers. The form is as follows:

$$XX.XXX \, \overline{)\, .YYY\,YYY}$$

Examples	$.25\,\overline{)\,1.50}$	$-1.50/.25$	$.75/-.005$	$.25 \div 1.5$
Rewritten	$25\,\overline{)\,150.0}$	$25\,\overline{)-150.0}$	$-5\,\overline{)750.0}$	$15\,\overline{)2.5}$
Answers	6	-6	-150	$.17$

To divide a fraction by a whole number, multiply the denominator of the fraction by that number.

Examples	$\dfrac{2}{7} \div 10$	$\dfrac{2}{7} \div -10$	$\dfrac{-2}{7} \div 10$
Rewritten	$\dfrac{2}{(7 \times 10)}$	$\dfrac{2}{(7 \times -10)}$	$\dfrac{-2}{(7 \times 10)}$
Answers	$\dfrac{2}{70}$	$-\dfrac{2}{70}$	$-\dfrac{2}{70}$

To divide two fractions, do the following steps: 1) invert the divisor, that is, make the original numerator the new denominator and the original denominator the new numerator; 2) multiply the denominators of the dividend and inverted divisor; and 3) multiply the numerators of the dividend and inverted divisor.

Examples	$(^3/_5) \div (^2/_5)$	$(-^3/_5) \div (^2/_5)$	$(^3/_5) \div (-^2/_5)$
Invert divisor	$(^5/_2)$	$(^5/_2)$	$(^5/-_2)$

Multiply	$\dfrac{(3 \times 5)}{(5 \times 2)}$	$\dfrac{(-3 \times 5)}{(5 \times 2)}$	$\dfrac{(3 \times 5)}{(5 \times -2)}$
Answers	$\dfrac{15}{10}$	$-\dfrac{15}{10}$	$-\dfrac{15}{10}$

2.5 EXPONENTS

Exponents indicate how many times a number, called a *base,* should be multiplied by itself. Exponents are indicated by a small number placed above and to the right of the base number.

Examples	3^1	3^2	3^3	3^4
Answers	3	$3 \times 3 = 9$	$3 \times 3 \times 3 = 27$	$3 \times 3 \times 3 \times 3 = 81$

Mathematicians have agreed that any number with an exponent of zero is equivalent to 1; the only exception is zero itself, which always remains zero.

Examples	3^0	4^0	5^0	0^0	$(^1/_2)^0$
Answers	1	1	1	0	1

2.6 SQUARE ROOTS

The symbol for square roots is the radical sign, which is written as $\sqrt[2]{}$ or $\sqrt{}$. To obtain a square root, find the number which when multiplied by itself (or squared) results in the number under the square root sign. Square roots may generally only be done with positive numbers.

This book assumes that the student has a calculator available to solve square roots, so there is no need to learn how to calculate them. To refresh your memory of square roots:

Examples	$\sqrt{1}$	$\sqrt{4}$	$\sqrt{9}$	$\sqrt{.25}$	$\sqrt{^1/_{36}}$	$\sqrt{49^2}$
Answers	1	2	3	.5	$^1/_6$	49

2.7 ABSOLUTE VALUES

The absolute value of a number is always the positive value of that number. The absolute value sign is two vertical lines on either side of the number.

Examples	$\mid 5 \mid$	$\mid -5 \mid$	$\mid {}^1/_5 \mid$	$\mid -{}^1/_5 \mid$	$\mid .5 \mid$	$\mid -.5 \mid$
Answers	5	5	$^1/_5$	$^1/_5$	.5	.5

2.8   ROUNDING AND TRUE CLASS LIMITS

In mathematics, it often happens that the answer to a problem is a number with several decimal spaces. The question is then "How many decimal spaces should be reported?" The rules are:
1.   While analyzing data, keep all decimal spaces.
2.   When reporting data, keep *no more* than two decimal spaces unless more are really needed. When there are more decimal spaces than needed:
   a)   Round off the unnecessary numbers, italicized in the examples below, by changing the last number to be kept to the next *nearest* whole number.
     6.25*6* becomes 6.26
     6.25*3* becomes 6.25
     6.*1* becomes 6
     6.*8* becomes 7
   b)   If the number to be rounded off is halfway between two whole numbers, round to the next nearest even number:
     6.255 becomes 6.26
     6.5 becomes 6
     5.5 becomes 6

These rounding rules are based on the concept of *true class limits.* This concept means that every number is conceived of as an interval rather than a point. For example:

| | True class limits of a number | | |
Number	Lower		Upper
−2	−2.5	to	−1.5
−1	−1.5	to	−.5
0	−.5	to	.5
1	.5	to	1.5
2	1.5	to	2.5

A number, then, represents an interval bounded by lower and upper class limits. There is a range of points within this interval, and these points are assumed to be equally distributed within the interval. The rounding rules assure that an observed number is assigned to its correct or true class interval. Note, however, that the upper limit of one interval is the lower limit for the next interval. To place a number correctly, it is necessary to have it one decimal place longer than the interval. If the number is only 1.5, for instance, it may not be correctly placed; however, if it is 1.56, then it may always be correctly placed in the "2" interval.

Chapter 3

Review of
Statistical Notation

This chapter reviews the basic notation used in statistics. More notation is introduced later on, but it will be relatively simple in comparison.

3.1 EQUALITY AND INEQUALITY

The equals sign ($=$) generally indicates that the two sides of an equation are equal. For example:

$$2 + 2 = 4.$$

The left side of the equation, then, has the same numerical value as the right side of the equation.

Sometimes, however, the equals sign is used to indicate the probability of a specific outcome. In flipping a coin, for example, the probability of getting heads is 50 percent:

$$\text{probability (heads)} = .50.$$

The context in which the equals sign is used should make it clear whether it refers to a certain result or a probable outcome.

When an equals sign has a slash through it ($\neq$), it means "does not equal." Because 2 does not equal 4, this could be written as $2 \neq 4$.

3.2 OPERATORS

Operators express either what should be done to obtain a result or what should be the relationship between groups.

Arithmetic operators:	$+$	Add
	$-$	Subtract
	$\times$, $\bullet$ or	
	$(\ )(\ )$	Multiply
	$\div$ or $/$	Divide
Relationship operators:	$A > B$	A is greater than B
	$A \geq B$	A is greater than or equal to B
	$A < B$	A is less than B
	$A \leq B$	A is less than or equal to B

3.3 PARENTHESES

Parentheses are used in several ways:

First, parentheses may be used to indicate multiplication. For example, 2×3 may be shown as (2) (3).

Second, parentheses may be used to show what should be considered as separate terms. For example, $2 + 3$ may be written as $(2) + (3)$.

Third, parentheses may be used to show the order in which computations should be done. Always calculate what is in the innermost parentheses first and then work outwards. For example:

$$((4 \times 5) \div (2 + 3)) =$$
$$(20 \div 5) = 4.$$

3.4 FACTORIALS

Factorials refer to the multiplication of a number with all numbers below it. For example, "five factorial" means:

$$5 \times 4 \times 3 \times 2 \times 1.$$

The symbol for this procedure is an exclamation sign (!) after the highest number of the factorial:

$$5! = 5 \times 4 \times 3 \times 2 \times 1.$$

3.5 SUMMATION NOTATION

It is often necessary to add lists of numbers in computing statistics. Rather than always writing out the entire list, a special operator or symbol is used to indicate which numbers are to be summed. The symbol is the Greek capital sigma, Σ.

Suppose, for example, that someone wants to know the number of children that the 10 welfare mothers have in Table 3.1. The first column identifies each mother or research subject by a number 1 through 10. The second column lists the actual number of children each mother has. Mother 1 has 4 children, mother 2 has 3 children, mother 3 has 2 children, and so on. The total number of children is obtained by adding all the numbers in the second column, $4 + 3 + 2 + 4 + 5 + 1 + 10 + 2 + 3 + 3 = 37$ children.

Because it is cumbersome to write out the list of numbers to be added, a conceptual shorthand is used:

$$\sum_{i=1}^{n=10} Y_i$$

Table 3.1. Identification numbers for welfare mothers and their number of children

Welfare mothers	Number of children
1	4
2	3
3	2
4	4
5	5
6	1
7	10
8	2
9	3
10	3
	37

This notation is read as "the summation of Y values from $i = 1$ through 10." To understand the use of this notation, several steps must be explained.

First, the symbol i is used to represent all identification numbers for all research subjects, ranging from 1 for the first subject through n for the last subject. The symbol n is generally used to indicate the last subject in the total list of subjects. For the 10 welfare mothers, the i column of Table 3.2 shows that i represents welfare mothers labeled 1 through 10 or 1 through n.

Table 3.2. Notation for identification numbers of welfare mothers and their number of children

Welfare mothers (i)	Number of children (Y)
1	Y_1
2	Y_2
3	Y_3
4	Y_4
5	Y_5
6	Y_6
7	Y_7
8	Y_8
9	Y_9
$n = 10$	Y_{10}
	$\sum_{i=1}^{n=10} Y_i$

Second, an alphabet symbol such as Y is used to represent the values to be added, or, in this example, the number of children. The Y column in Table 3.2 refers to the number of children for all mothers. Each Y in the column, however, is followed by a subscript to indicate the number of children a specific subject has. Thus, Y_1 (read as Y-sub-one) refers to the four children that the first welfare mother has, Y_2 (read Y-sub-two) refers to the three children that the second welfare mother has, Y_3 (read as Y-sub-three) refers to the two children that the third welfare mother has, and so on.

Third, the symbol Σ is used to represent the addition or summation of a list of numbers. The subscript below it ($\Sigma_{i=1}$) indicates the first subject whose value is to be added in the summation sequence. The number above it (Σ^n), indicates the last subject whose value is to be included in the summation. In the welfare mother example, mothers 1 through 10 should be included, or $\Sigma_{i=1}^{10}$.

Lastly, the complete summation notation consists of both the summation sign, with numbers below and above to indicate which subjects are to be included in the summation, and the correct alphabet symbol with its subscript i to indicate what values are to be added. Thus, $\Sigma_{i=1}^{10} Y_i$, or $\Sigma_{i=1}^{n} Y_i$, is the complete summation notation needed to refer to the total of 37 children belonging to the 10 welfare mothers.

Now use Table 3.1 to add the following numbers and to practice the use of summation notation:

1. sum the number of children that the first five welfare mothers have;
2. sum the number of children that the third through sixth welfare mothers have; and
3. sum the number of children that the fifth through ninth welfare mothers have.

The answers are:

1) $\displaystyle\sum_{i=1}^{5} = Y_1 + Y_2 + Y_3 + Y_4 + Y_5$

$= 4 + 3 + 2 + 4 + 5$
$= 18$ children

2) $\displaystyle\sum_{i=3}^{6} = Y_3 + Y_4 + Y_5 + Y_6$

$= 2 + 4 + 5 + 1$
$= 12$ children

3) $\displaystyle\sum_{i=5}^{9} = Y_5 + Y_6 + Y_7 + Y_8 + Y_9$

$= 5 + 1 + 10 + 2 + 3$
$= 21$ children.

3.6 ABBREVIATION OF SUMMATION NOTATION

Because in the calculation of statistics it is frequently necessary to sum score values for all subjects, the numbers below and above the summation sign are often dropped to become:

$$\Sigma Y_i$$

Note that the subscript i after the Y has been retained. This indicates that i is to take on the successive values of subject 1, 2, 3, up to n. The reason for this abbreviation is simply that it is easier and faster to write. This book will always use the abbreviated form.

3.7 PARENTHESES AND SUMMATION NOTATION

When there are no parentheses, the summation sign includes only those symbols that stand immediately next to the summation sign *and* that are not separated by a plus or minus sign. When a plus or minus sign comes between, however, the summation sign does not include anything after the plus or minus. For example, $\Sigma Y_i + 10$ instructs you to add all the Y_i values first and only afterwards add 10 to that total sum.

When there are parentheses, however, the above rule changes. The summation sign now includes all those symbols and numbers enclosed within the parentheses. Thus, for example, $\Sigma(Y_i + 10)$ now instructs you to add 10 to each Y_i value and then add them all together.

Lastly, the location of parentheses is important. ΣY_i^2, $\Sigma(Y_i^2)$, and $\Sigma(Y_i)^2$ each tell you to square every Y value first and then add all squared values together, but $(\Sigma Y_i)^2$ tells you to sum all Y values first and then square the total sum. Using the numbers in Table 3.3, the first three summations result in 53, whereas the last summation results in 121, or 11^2.

3.8 DOUBLE SUMMATION NOTATION

Often, in the calculation of statistics, it is necessary to add lists for more than one group at the same time. Double summation notation provides a shorthand method to indicate the necessary addition.

Table 3.3.   Numerical example

i	Y	Y^2
1	4	16
2	6	36
3	1	1
	11	53

Suppose, for example, that someone wants to know the number of children that welfare mothers have. However, instead of just one group of mothers, there are three groups: those on welfare for less than a year, from 1 to 5 years, and over 5 years. Suppose there are five mothers in the first group, six mothers in the second group, and seven mothers in the third group. Table 3.4 shows the notation and the actual number of children for each group and the total sample. Double notation summarizes Table 3.4 as follows:

$$\sum_{j=1}^{k} \sum_{i=1}^{n_k} Y_{ij} = \Sigma Y_{i1} + Y_{i2} + \Sigma Y_{i3}$$
$$= 9 + 22 + 14$$
$$= 45.$$

To understand this notation, several steps must be explained.

First, the symbol j is used to represent group membership and ranges from 1 (the first group) to k (the last group). Second, the symbol i is used to represent the number of subjects in each group and ranges from 1 (the first subject in the group) to n_k (the last subject in the group). The letter n without a subscript refers to the total number of subjects within each group. Third, the symbol Y represents the values to be added, or, in this example, the number of children. Thus, a Y_{ij} represents the score value (Y) for the ith subject in the jth group. Y_{11}, for example, indicates the number of children (one) that the first mother in the first group has, a Y_{63} indicates the number of children (two) that the sixth mother in the third group has, and so on.

Summation signs attached to these symbols specify which values should be added. For example, ΣY_{i1} says to add the score values for all subjects in group 1. Their total is nine children. The same notation is used

Table 3.4. Notation and values for number of children of three groups of welfare mothers

Group I		Group II		Group III	
Y_{11}	1	Y_{12}	3	Y_{13}	1
Y_{21}	2	Y_{22}	4	Y_{23}	2
Y_{31}	3	Y_{32}	5	Y_{33}	3
Y_{41}	2	Y_{42}	4	Y_{43}	2
Y_{51}	1	Y_{52}	3	Y_{53}	1
		Y_{62}	3	Y_{63}	2
				Y_{73}	3
	$\Sigma Y_{i1} = 9$		$\Sigma Y_{i2} = 22$		$\Sigma Y_{i3} = 14$
		$\Sigma\Sigma Y_{ij} = 45$			

to add the score values for each group. If there is a double summation sign such as

$$\sum_{j=1}^{k} \sum_{i'=1}^{n_k} Y_{ij}$$

first sum all of the score values within each group ($\Sigma Y_{i1} + \Sigma Y_{i2} + \Sigma Y_{i3} = 9 + 22 + 14$) as noted by the summation sign with the subscript i, and then sum the group totals together (45) as noted by the summation sign with the subscript j. The result is usually called the *grand total* to distinguish it from the separate group totals.

PART II

DESCRIPTIVE STATISTICS

Administrators, program staff, and social scientists often have a large amount of information about clients, staff, and programs. How to make this mass of information useful is an ongoing challenge.

The purpose of *descriptive statistics* is to organize and summarize information so it may be easily used. When a local welfare office prepares an annual report, for example, descriptive statistics help to organize and summarize client, staff, and cost information on its programs. Or when a therapist wants to summarize client progress in several treatment programs, descriptive statistics about the client are used to organize the information for purposes of assessment.

Before discussing types of descriptive statistics and their uses, two clarifications in terminology are necessary. First, the usual word for information in statistics is *data*. The information you have is usually referred to as data. Second, data refers to subjects and their descriptive variables. *Subjects* are the things, programs, or persons about which data have been collected. Subjects are the focus of description. *Variables* are descriptive characteristics of subjects that vary or take on different values or scores across subjects. Thus, in the local welfare office example above, the subjects are its programs and the variables are the clients, staff, and cost characteristics within each program. In the therapist example, the subject is the client and the variables are client behaviors or other characteristics expected to change in each treatment program. Variables, then, are used to describe subjects.

The types of statistics defined and used in the following chapters of Part II are listed below.

Data Lists These describe each subject across all variables. Each client, for example, has scores on different diagnostic tests. The scores for each client are presented in a data list. Data lists are discussed in Chapter 4.

Frequency Distributions These describe each variable across all subjects. Each diagnostic test, for example, has a range of scores. A frequency distribution shows how often each score occurs across all subjects. All subjects' scores for each variable may be summarized by distributions of frequencies, proportions, percentages, or percentiles. Frequency distributions are discussed in Chapter 5.

19

Measures of Central Tendency These use a single number to describe the "typical" score among all subjects' scores on a variable. The average (mean) age of subjects is one common example. The mean, median, and mode are such single numbers. Measures of central tendency are discussed in Chapter 7.

Measures of Variability These use a single number to describe the spread or scatter among all subjects' scores on a variable. If all scores on an intelligence test are similar, there is little variability among subjects' scores; if all scores are very different, there is a large variability among subjects' scores. The variance, standard deviation, range, and variation ratio are such single numbers. Measures of variability are discussed in Chapter 8.

Bivariate Frequency Distributions These describe the joint distribution of two variables by showing the frequency with which the scores of both variables occur together across all subjects. If sex and income were two variables used to describe a group of adult subjects, for example, a bivariate frequency distribution might show that more males than females had a high income and more females than males had a low income. Bivariate frequency distributions are discussed in Chapter 9.

Measures of Relationship These use a single number to describe the strength of relationship or association between two or more variables. If program budget size increases as number of clients increase, then the measure of relationship will indicate how strong their relationship is. Some of the measures of relationship discussed in Chapter 11 are lambda (λ), gamma (γ), Pearson r, and r squared (r^2).

Graphs These are pictures that may be used to highlight visually the important characteristics of both univariate and bivariate distributions. Graphs for univariate distributions are discussed in Chapter 6, and graphs for bivariate distributions are discussed in Chapter 10.

Choice among these types of descriptive statistics depends largely on your notion of what is most appropriate in your situation. Figure 1, however, suggests the broad outlines for your decision making. To read this figure, start at the upper left box where it asks if you want to "Stay with your variables as originally measured?" If your answer is yes, follow the YES arrow. If your answer is no, follow the NO arrow. Follow this yes-or-no procedure until you reach a decision box without any more yes or no arrows (usually the box on the far right side of the page). The decision box you reach tells you what chapter you should read to accomplish the suggested techniques.

Although Figure 1 provides an overall decision-making method, any or all techniques may be used at once. The purpose of the chart is simply to help you make a rational choice. Moreover, once the techniques in the following chapters are mastered, Figure 1 will become more useful as a quick summary of descriptive techniques and where to find a detailed discussion of them in this book.

Chapters 4 through 11 now introduce these techniques. Chapter 12 shows how to choose and combine these techniques for a research report.

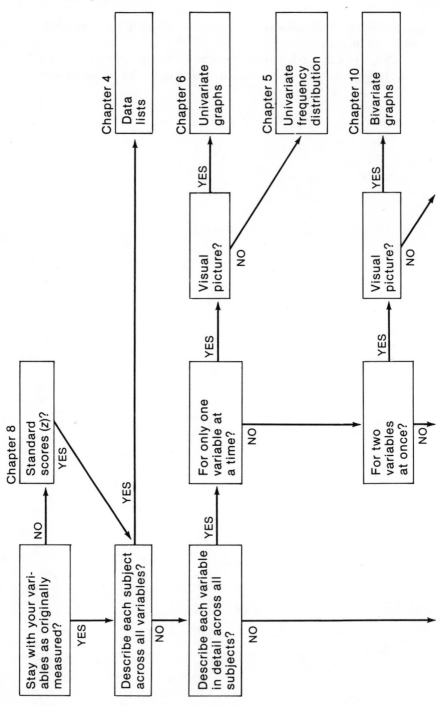

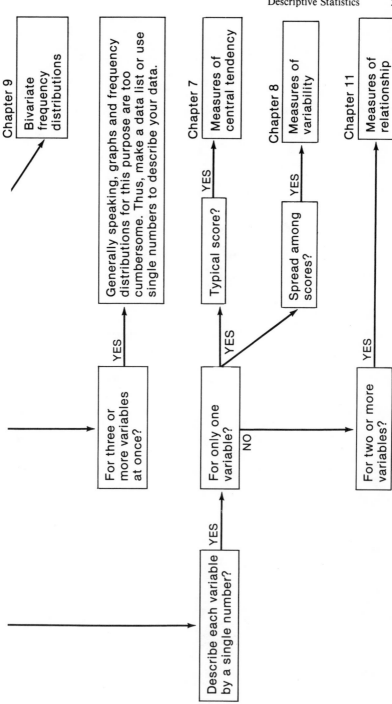

Figure 1.   Decision-making chart for choosing an appropriate type of descriptive analysis

Chapter 4

Data Lists and SPSS

One way to organize information in descriptive statistics is to make a data list.

4.1 DEFINITION

A *data list* describes each subject across all variables by making a list of subjects and their scores on each variable. Each row of a data list represents one subject and that subject's scores on every variable. Each column of a data list represents one variable and the scores for all subjects on that variable.

4.2 EXAMPLE

An example of a data list is given in Table 4.1. The title of this table specifies what type of table it is (a data list), who the subjects are (girls in a detention home), and the number of subjects ($n = 30$). All data lists should have a title, and the content of the title should follow this format to tell readers what the table is about.

There are four substantive variables in this data list: age, religion, the number of days spent in the detention home, and seriousness of offense. Note that information in a data list may be scaled as a nominal variable (religion), an ordinal variable (seriousness of offense), or an interval variable (age or number of days spent in the detention home). The scale values for nominal and ordinal variables always need verbal definitions to be understood. These definitions are often shown at the bottom of data tables, as in Table 4.1.

Information about each variable alone is found by reading *down a row*. Therefore, the column with the heading "subjects' identification number" is usually referred to as the row variable because you read across each row to get the relevant information about a single subject. The 15th subject (or 15th row) in this data list, for instance, is 11 years old, Catholic, spent 1 day in the detention home, and committed an offense of little or low seriousness.

25

Table 4.1.   A data list for girls in a detention home ($n = 30$)

Subject identification number	Age	Religion[a]	Number of days spent in detention home	Seriousness of offense[b]
1	11	1	5	1
2	17	3	11	3
3	14	1	8	1
4	13	3	5	1
5	13	2	10	3
6	14	2	9	3
7	14	2	12	2
8	16	2	13	3
9	13	1	7	1
10	12	1	4	1
11	15	3	3	2
12	10	1	8	2
13	15	3	9	2
14	14	1	5	2
15	11	2	1	1
16	12	2	6	1
17	14	1	9	2
18	15	1	2	3
19	16	1	14	3
20	13	2	9	1
21	15	2	4	3
22	14	1	11	3
23	17	1	8	3
24	16	2	7	2
25	12	1	4	2
26	13	1	7	1
27	13	2	9	1
28	12	2	6	1
29	11	1	5	1
30	10	1	4	2

[a]Religion has three categories: 1 = Protestant, 2 = Catholic, and 3 = other.

[b]Seriousness of offense has three categories: 1 = low seriousness, 2 = moderate seriousness, and 3 = high seriousness.

4.3   USE

As demonstrated by Table 4.1, data lists contain a lot of information that is relatively difficult to assimilate quickly. Thus, if used for applied purposes, data lists are only appropriate for providing periodic census reports about subjects that must include information on a subject-by-subject basis. Otherwise, the main purpose of data lists is to compile information for data processing on a computer as in management information systems or in statistical analysis.

4.4   DATA PROCESSING AND THE COMPUTER

Most data processing is done by the computer because it is much easier and faster than hand calculations. Since most government agencies and universities have computer facilities, each chapter of this book henceforth demonstrates how to use the computer to calculate the statistical procedures presented.

There are three chief problems in the use of computers: how to get data ready for computer analysis, how to use computer programs to analyze data, and how to interpret the statistical results. The first problem will be discussed immediately. The second problem is introduced here, and then also discussed at the end of each chapter. The third problem is discussed with the presentation of every statistic in this book.

One typical way to get data into the computer is to use computer punchcards. Computer cards are designed to have up to 80 numbers or letters punched into them at a keypunch, a machine that operates just like a typewriter. The computer than reads the punched holes on the cards to store your data in its central memory. There is no limit on the number of cards that may be used.

If computer cards were punched for all the girls in the detention home (see Table 4.1), there would be a total of 30 data cards, or one card for each subject. Each card corresponds exactly to one row on the data list, with the subject's identification punched in first, and then the subject's age, religion, number of days spent in the detention home, and seriousness of offense. Although the order of variables may be changed, this was the order chosen for this example. The important point is that the order of variables, once chosen, must remain consistent on all cards.

To punch cards correctly requires that every subject's identification number *and* scores on the substantive variables *always* be placed in the same column(s) on each computer card. The initial choice of what columns should represent which variables is open to any decision, but thereafter complete consistency with that decision is essential. For the detention home data in Table 4.1, the decision was made to punch the subject's identification number into columns 1 and 2 of the computer card (two columns are needed because the identification number may take two spaces), age into columns 3 and 4, religion into column 5, number of days spent in the detention home into columns 6 and 7, and seriousness of offense into column 8. Once this decision was made, the data on each of the 30 cards were punched in exactly the same way.

Correct punching of computer cards also requires that numbers are "right-justified." This means that the last digit in a number must be punched in the rightmost column of its assigned space on the computer card. For example, the subject's identification number is assigned columns 1 and 2. If the identification number is 1, this should be punched as "01." The "1" is thus right-justified. If the identification number is 15, this should be punched as "15" so that the last digit, 5, is right-justified.

An alternative to computer cards is magnetic tape. Data and programs are entered on the magnetic tape by cards or by remote terminals. An "interactive mode" rather than a "batch mode" is then used to instruct the computer. However, this book will refer only to cards and the batch mode because it is a universal procedure.

The procedure used to summarize the organization of data is called a *codebook*. A codebook provides written documentation of how data are organized on computer card for computer analysis. Table 4.2 is an example of a codebook for the detention home study. The first vertical section of this codebook refers to the variable's column location on the computer cards. The second vertical section refers to the variable's name used in the actual computer analysis. This will be discussed in Section 4.6. The last vertical section of a codebook includes a description of each variable and its scores or codes.

Once the codebook and card punching are done, the next problem is how to use computer programs to analyze your data. The usual procedure in the social sciences is to use "canned" or prepackaged programs already stored in the computer. There are many canned programs, each of which will analyze your data once you call forth the correct procedures. Procedures are called forth by using code words or numbers, much like telephone numbers are used to call a friend. Each set of canned programs uses different code words; thus, to simplify matters, this book only uses one

Table 4.2.   Codebook for the detention home study

Column	Variable name	Variable description
1–2	SUBJECT	Subject's identification number (01 to 30)
3–4	AGE	Age of subject in years
5	RELIGION	Religion 1 = Protestant 2 = Catholic 3 = other
6–7	DAYS	Number of days spent in the detention home
8	OFFENSE	Seriousness of offense before admission to detention home 1 = low seriousness 2 = moderate seriousness 3 = high seriousness

canned program called the *Statistical Package for the Social Sciences,* or SPSS.

4.5 INTRODUCTION TO SPSS

SPSS is a set of computer programs that provides many different types of statistics with which to analyze data. There are two advantages to using SPSS. First, SPSS is available in most computer facilities in the United States and in many around the world. Once SPSS is learned, this availability means that anyone may analyze data anywhere with relative ease. Second, the published manuals on how to use SPSS are written extremely well. These manuals use plain English to describe the use of each SPSS program, provide well documented examples, and even review the conceptual base for each statistic and how to interpret them.

There are currently three different SPSS manuals available:

1. *Statistical Package for the Social Sciences, Second Edition,* by N. H. Nie, C. H. Hull, J. G. Jenkins, K. Steinbrenner, and D. E. Bent. New York: McGraw-Hill, 1975.
2. *SPSS Primer* by W. R. Klecka, N. H. Nie, and C. H. Hull. New York: McGraw-Hill, 1975.
3. *SPSS Update 7-9* by C. H. Hull and N. H. Nie. New York: McGraw-Hill, 1981.

The first manual, the *Statistical Package for the Social Sciences,* is the basic manual for Version 7.0 SPSS. There have been six previous versions. This manual is about 700 pages long and describes in detail how to use SPSS and how to analyze data with the most widely used statistics. The second manual, the *SPSS Primer,* is an introductory text that discusses the fundamentals of SPSS and the simpler statistics in about 140 pages. Its advantage is its relative simplicity and lower cost. The last manual, the *SPSS Update 7-9,* presents statistics not originally available in the first manual (new nonparametric tests and reliability tests) and corrections and improvements on it.

This book will refer to all three manuals to show how to use SPSS to analyze the data presented here. Whenever possible, however, emphasis will be placed on the *SPSS Primer.* Thus, it will be assumed that each student owns the *SPSS Primer* and that the school library will have the other two manuals as reference material. The end of every chapter of this book will have an SPSS section that demonstrates correct computing procedures. The following section on SPSS will begin this process by explaining how to arrange data for SPSS computer analysis. Before doing so, read Chapters 1, 2, and 3 of the *SPSS Primer* as supporting material for Section 4.4, and then read Chapters 4 and 5 of the *SPSS Primer* as supporting material for Section 4.6.

4.6 SPSS

For the SPSS computer programs to read and analyze data, more computer cards must be punched. These cards contain information in columns 1 through 15 (called the *control field* in SPSS) and in columns 16 through 80 (called the *specification field* in SPSS). The following examples identify what belongs in the control field, columns 1 through 15, by placing the numeral 1 at the top of the sequence, and what belongs in the specification field, columns 16 through 80, by placing the numeral 16 at the top of the sequence.

It is assumed that the *SPSS Primer* will be read concurrently with the provided examples, so lengthy explanations are not given. For this section, read Chapters 4 and 5 in the *SPSS Primer*.

The following SPSS cards are required to enter data cards into the computer:

Columns 1–15 Control field	Columns 16–80 Specification field
1	16
DATA LIST	FIXED (1)/1 SUBJECT 1–2, AGE 3–4, RELIGION 5, DAYS 6–7, OFFENSE 8
INPUT MEDIUM	CARD
N OF CASES	30

The DATA LIST card defines the organization of data in a way that allows SPSS to use it. DATA LIST is punched in the control field. At the beginning of the specification field, FIXED is punched in to indicate that each variable has an unchanging or fixed column location on the computer data cards. The number of cards used for each subject or case is then placed within parentheses. Each subject in the detention home study has only one card, so a 1 is placed within parentheses. The parentheses must be followed by a slash. The remainder of the DATA LIST card contains what is called a *variable list*. The variable list identifies what card the variables are located on, their name, and column location. The numeral 1 after the slash indicates that the variables listed after it are on the first and only card for each subject. A space is then left and the first variable name is punched in along with its column location. Each variable name must start with an alphabet letter and may be no more than eight characters in length. I have chosen variable names that would remind me of what each variable was describing: SUBJECT for subject's identification number, AGE for age, RELIGION for religion, DAYS for the number of days spent in the detention home, and OFFENSE for seriousness of offense. Since SUBJECT is located in the first two columns, the numerals 1-2 are punched in and followed by a comma. This process is done for each variable until the last variable is done. No comma follows the last column number referenced on the variable list. If more than one card is needed to punch in the variable list, use as many cards as necessary but always start in column 16.

The INPUT MEDIUM card tells the computer that the data are on computer cards rather than on tape or disc. The N OF CASES card tells the computer that this data set has 30 cases or subjects.

Other SPSS cards may be put after the above cards as well, but they are optional. The MISSING VALUES card tells the computer what code number is used for missing data (for example, a zero or a blank in the column for religion would indicate that the researcher did not know that subject's religion). Since there are no missing data in the detention home data list, this card has not been used.

Another optional SPSS card is the VALUE LABELS card. This card tells the computer to print the name of each variable score interval where appropriate. The VALUE LABELS card is especially useful for labeling the score values of nominal and ordinal variables. In the detention home study, for example, the VALUE LABELS card is used to label what the numbers 1, 2, and 3 mean for religion, a nominal variable, and seriousness of offense, an ordinal variable. For religion, 1 means Protestant, 2 means Catholic, and 3 means all other religions. For seriousness of offense, 1 means an offense of low seriousness, 2 means an offense of moderate seriousness, and 3 means an offense of high seriousness.

Another optional SPSS card is the VAR LABELS card. This card is useful if the whole variable name should be printed out rather than just its name on the DATA LIST. For example, instead of just printing the name OFFENSE, use of the VAR LABELS card would print "seriousness of offense" instead.

Shown below is how these optional SPSS cards are used in the detention home study:

1: Control field	16: Specification field
VALUE LABELS	RELIGION (1) PROTESTANT (2) CATHOLIC (3) OTHER/
	OFFENSE (1) LOW (2) MODERATE (3) HIGH/
VAR LABELS	SUBJECT, SUBJECT IDENTIFICATION NUMBER/
	AGE, AGE AT TIME OF DETENTION/
	RELIGION, RELIGION/
	DAYS, NUMBER OF DAYS IN DETENTION/
	OFFENSE, SERIOUSNESS OF OFFENSE/

Each of the SPSS cards shown above should be punched as above on computer cards, if the program is to run without errors. If you wish to change the cards, check the *SPSS Primer* first to ensure you have done it correctly.

Chapter 5

Univariate
Frequency Distributions

Rather than provide all information on a single subject as in a data list, another way to organize information descriptively is to make a univariate (one-variable) frequency distribution for each variable.

5.1 DEFINITION

A univariate frequency distribution describes each variable by making a list of all possible scores in one column and the frequency with which they occur across subjects in a second column.

Since both data lists and frequency distributions organize information, the choice between them depends on how the information is to be used. If all information about a subject is desired, for example for periodic census reports, a data list may be preferable. If specific information about a single variable is required, for example the number of days spent in the detention home, a univariate frequency distribution may be more appropriate. Generally, however, data lists are used to compile information for data processing on a computer, and frequency distributions are used for descriptive data analysis in research reports and public presentations.

Once a choice is made to use univariate frequency distributions, additional decisions need to be made whether to use ungrouped or grouped scores and whether to report frequencies, proportions, percentages, or percentile ranks across subjects. These topics and a few more are covered in this chapter. Figure 5.1 reflects these choices and suggests some ways to proceed.

5.2 DISTRIBUTIONS WITH UNGROUPED SCORES

A univariate frequency distribution with ungrouped scores lists *all possible scores* in the first column and their *frequency* or number of times each score occurs across subjects in the second column. The title of the first column, where the scores are listed, should be the name of the variable.

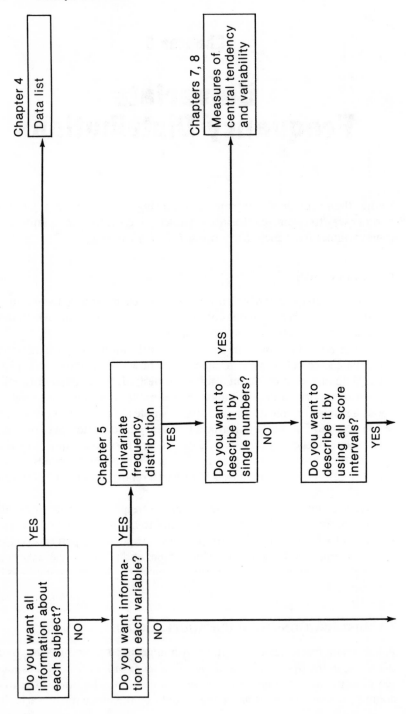

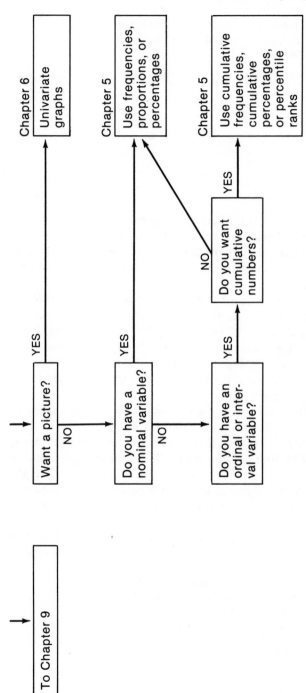

Figure 5.1. Decision-making chart for describing univariate frequency distributions.

Table 5.1. Frequency distribution of age for the girls in the detention home ($n = 30$)

Age	Frequency (f)
10	2
11	3
12	4
13	6
14	6
15	4
16	3
17	2

The title of the second column, where the frequency of each score is listed, should be "frequency" or its symbol "f." When ordinal or interval variables are used, it is customary to order the scores from low to high in the first column. When nominal variables are used, however, the scores may be put in any order.

Using the information from the detention home data list in Table 4.1, Table 5.1 shows the univariate frequency distribution for the interval variable called "age." Table 5.2 shows the frequency distribution for the nominal variable called "religion." Table 5.3 shows the frequency distribution for the interval variable called number of "days spent" in the detention home. Table 5.4 shows the frequency distribution for the ordinal variable called "seriousness of offense."

The data in Table 5.1 can be used to show how to read a frequency distribution. Two girls (frequency of subjects) are age 10 (score) in this detention home, three girls are age 11, four girls are age 12, six girls are age 13, six girls are age 14, four girls are age 15, three girls are age 16, and two girls are age 17. If one looks at the whole distribution, most girls (frequency of $20 = 4 + 6 + 6 + 4$) in this detention home are between the ages of 12 and 15. Few girls are younger (10 or 11 years old) or older (16 or 17 years old).

Notice that the total of the frequency column in Table 5.1 is 30, the total number of girls in the detention home. A total of less than 30 would

Table 5.2. Frequency distribution of religion for the girls in the detention home ($n = 30$)

Religion	Frequency (f)
Protestant	15
Catholic	11
Other	4

Table 5.3. Frequency distribution of days spent by the girls in the detention home ($n = 30$)

Days spent in detention home	Frequency (f)
1	1
2	1
3	1
4	4
5	4
6	2
7	3
8	3
9	5
10	1
11	2
12	1
13	1
14	1

indicate that there is an addition error or some missing data (uncollected or unavailable information). The total of the frequency column should always equal the total group size ($n = 30$) given in the title of the table.

Also notice that the title of Table 5.1 specifies what type of table it is (a frequency distribution with "univariate" assumed unless stated otherwise), the variable in the table (age), the subjects (girls in a detention home), and the number of subjects ($n = 30$). All frequency distribution tables should have a title, and the content of the title should follow this format to tell readers what the table is about.

5.3 DISTRIBUTIONS WITH GROUPED SCORES

When a variable has many possible score values, for example, 10 or more, the scores are often grouped together in *categories* or *score intervals* to make a frequency distribution easier to examine. Whereas the size of the

Table 5.4. Frequency distribution of seriousness of offense for the girls in the detention home ($n = 30$)

Seriousness of offense	Frequency (f)
Low	12
Moderate	9
High	9

score interval is solely up to the discretion of the user, the width of each interval in the table should be the same. As always, there are exceptions to the rule. One exception is the occurrence of open-ended intervals at the beginning or end of a distribution. Open-ended intervals occur when there are few observations at the extreme ends of a distribution. For example, few people age 18 and below or 65 and above receive unemployment insurance. Thus, in this case it is more efficient to use open-ended intervals such as people age 12 and below or age 65 and over than a long list of intervals of equal size that contain few or no subjects. The second exception is grouping score intervals for nominal variables, where the notion of equal interval size is meaningless. For nominal scores, common sense or some conceptual reason are the main considerations. Instead of listing every possible religion, for example, religions may be grouped under the broader categories of Protestant, Catholic, or other as was done in Table 5.2.

Given these exceptions, Tables 5.5. and 5.6 demonstrate two ways of grouping score intervals for the variable called "days spent" in the detention home. The interval size in Table 5.5 is 2 days instead of the original 1-day interval in Table 5.4, and the interval size in Table 5.6 is 1 week.

The choice between using smaller or larger intervals depends on the amount of information to be conveyed. Generally, the smaller the interval size, the greater the information conveyed. To demonstrate this point, Tables 5.4, 5.5, and 5.6 show that about an equal number of girls spend 1 or 2 weeks at the detention home, but only Table 5.6 highlights this information. Yet Table 5.4 with the ungrouped 1-day intervals and Table 5.5 with the grouped 2-day intervals show additionally that most girls spend 5 to 10 days at the detention home and that few girls spend more than 10 days or less than 5 days. The additional information in these tables may therefore provide a better basis for program planning. The ad-

Table 5.5. Frequency distribution of days spent by the the girls in the detention home ($n = 30$)—An example of grouped score intervals with small width

Days spent in detention home	Frequency (f)
1–2	2
3–4	5
5–6	6
7–8	6
9–10	6
11–12	3
13–14	2

Table 5.6. Frequency distribution of days spent by the girls in the detention home $(n = 30)$—An example of grouped score intervals with large width

Days spent in detention home	Frequency (f)
1-7	16
8-14	14

ministrator and practitioner may decide to plan programs on a 5-day basis rather than a weekly basis, for example, to fit better the flow of girls into and out of the detention home. Since the grouped score intervals in Table 5.5 show the same important information as the ungrouped scores in Table 5.4, grouped score intervals may be preferred because they are easier to read.

Overall, then, smaller interval sizes are often more useful because they retain more information, but if the category intervals become too small, the message may be buried in too much detail—hence the original reason for using grouped scores. Knowing what data patterns are important, then, determines how small intervals should be.

From a statistical perspective, frequency distributions with grouped score intervals were once used to facilitate statistical calculations by hand. With the general availability and ease of using calculators and computers nowadays, however, only frequency distributions with ungrouped scores are used for data analysis to prevent loss of information. Thus, the major use of frequency distributions with grouped score intervals now is for such applied purposes as program planning or agency reports where no further data analysis needs to be done.

So far, univariate frequency distributions have been shown for reporting frequencies. However, such distributions may also be constructed using proportions, percentages, and percentile ranks.

5.4   PROPORTIONS

A *proportion* of a score interval, ungrouped or grouped, is defined as the frequency of a score interval divided by the total number of subjects:

$$p = \frac{f}{n}$$ where
p = proportion

f = frequency of a score interval

n = total number of subjects.

Proportions may be calculated for scores in nominal, ordinal, or interval scales.

The letter p actually denotes the proportion in a sample of subjects. The Greek letter π denotes the proportion in a population. As will be seen in Part III, these distinctions between symbols become important. You will also see that the letter p will be used to denote the probability of a score.

If the score intervals of a distribution are *mutually exclusive* (contain no overlapping score intervals) and *exhaustive* (contain all score intervals), the sum of proportions is always equal to 1.00, given rounding error. This occurs because proportions are calculated relative to the total number of subjects. In other words, the sum of all frequencies (Σf) is the same as the total number of subjects (n), and any number divided by itself equals 1.00 ($\Sigma f \div n = 1.00$, since $\Sigma f = n$, thus $n \div n = 1.00$). When proportions are calculated for each score interval, the result is called a *distribution of proportions*.

The third column in Table 5.7 shows how proportions are calculated for each interval of the girls' ages in the detention home. The proportion of girls who are 16 years old, for example, is .10. One tenth of the girls in this detention home are 16 years old.

From an applied perspective, proportions are useful in calculating percentages, which form the basis of most program and agency reports. From a statistical perspective, proportions are useful in understanding the basic concepts of probability and theoretical probability distributions, both of which form the basis of decision making in what is called inferential statistics (see Part III of this book).

5.5 PERCENTAGES

A *percentage* of a score, ungrouped or grouped, is defined as the proportion of a score interval multiplied by 100:

$$pc = p \times 100 \qquad \text{where} \quad pc = \text{percentage}$$
$$\text{and} \quad p = \text{proportion}.$$

Percentages, like proportions, may be calculated for scores in nominal, ordinal, or interval scales.

If the score intervals are mutually exclusive and exhaustive, the sum of percentages is always 100 percent. When percentages are calculated for each score interval, the result is called a *percentage distribution*.

The fourth column in Table 5.7 shows how percentages are calculated for each interval of the girls' ages in the detention home. The percentage of girls who are 16 years old, for example, is 10 percent. Ten percent of the girls in this detention home are 16 years old.

Table 5.7.   Frequencies, proportions, and percentages of age for girls in the detention home ($n = 30$)

Age	Frequency (f)	Proportion (p)	Percentage (pc)
10	2	$2 \div 30 = .07$	$.07 \times 100 = 7$
11	3	$3 \div 30 = .10$	$.10 \times 100 = 10$
12	4	$4 \div 30 = .13$	$.13 \times 100 = 13$
13	6	$6 \div 30 = .20$	$.20 \times 100 = 20$
14	6	$6 \div 30 = .20$	$.20 \times 100 = 20$
15	4	$4 \div 30 = .13$	$.13 \times 100 = 13$
16	3	$3 \div 30 = .10$	$.10 \times 100 = 10$
17	2	$2 \div 30 = .07$	$.07 \times 100 = 7$

Percentages are useful because they reflect the relative frequency with which scores occur rather than their absolute frequency. Percentages may only vary from zero to 100, whereas frequencies may range across any numbers. Percentages, then, provide a common reference point among all distributions—a characteristic that makes description and comparison much easier.

Comparisons of percentages may be made within a single distribution or between distributions. An example of comparisons within a single distribution is given in Table 5.7, which compares girls age 10 and 11 with girls age 12 and 13. The percentage column shows that there are 16 percent ($33\% - 17\% = 16\%$) fewer girls aged 10 and 11 ($7\% + 10\% = 17\%$) than there are girls at ages 12 and 13 ($13\% + 20\% = 33\%$). If frequencies were used, then the frequency column shows that there are five ($10 - 5 = 5$) fewer girls aged 10 and 11 ($2 + 3 = 5$) than there are at ages 12 and 13 ($4 + 6 = 10$). The 16 percent figure provides a better sense of the difference between these two age groups than the absolute number of 5.

An example of comparisons between distributions is given in Table 5.8, which compares the age distributions for girls in two different detention homes. Since the number of girls in each detention home differs, there being 30 girls in the first home and 60 girls in the second home, comparisons of percentages is more informative than comparisons of frequencies. Thus, although there are twice as many girls in the second detention home as in the first, the percentage distribution of age is the same for each home. It could not be said, therefore, that selection into these two homes is done discriminatorily on the basis of age.

The usefulness of percentages is sometimes overemphasized, however. One reason is that percentages and frequencies may tell different but equally truthful stories. Table 5.9 demonstrates this problem by comparing budget growth for two human services departments between 1975 and 1978. When absolute dollar figures are used, as in column 4, department 1 obviously has had a bigger budget increase ($\$1,200,000 - \$1,000,000 =$

Table 5.8.   Percentage distributions of age for girls
in two different detention homes

Age	Percentages for the first detention home ($n = 30$)		Percentages for the second detention home ($n = 60$)	
10	7	(2)^a	7	(4)
11	10	(3)	10	(6)
12	13	(4)	13	(8)
13	20	(6)	20	(12)
14	20	(6)	20	(12)
15	13	(4)	13	(8)
16	10	(3)	10	(6)
17	7	(2)	7	(4)

^{a}Frequencies are given in parentheses.

$200,000) than department 2 ($650,000 − $500,000 = $150,000). When percentage figures are used, however, department 2 obviously had had a relatively higher budget increase:

$$\frac{\text{budget increase}}{\text{1975 budget}} = \frac{\text{1978 budget} - \text{1975 budget}}{\text{1975 budget}}$$

$$= \frac{\$650,000 - \$500,000}{\$500,000}$$

$$= .30 \text{ or a 30 percent increase for department 2}$$

than department 1:

$$\frac{\text{budget increase}}{\text{1975 budget}} = \frac{\text{1978 budget} - \text{1975 budget}}{\text{1975 budget}}$$

$$= \frac{\$1,200,000 - \$1,000,000}{\$1,000,000}$$

$$= .20 \text{ or a 20 percent increase for department 1.}$$

Note that *rate increases* are calculated by taking the actual increase during a specified time period and dividing that increase by the amount at the beginning of the time period. *Rate decreases* are calculated in the same way: the actual decrease is divided by amount at the beginning of the time period. Because there is a decrease, the numerator will be negative and thus a minus sign will precede the rate to indicate a rate decrease.

Which department has had the largest budget increase? There can be no glib conclusions here because the decision depends on the purpose of

Table 5.9.   Total budget of two human service departments for 1975
and 1978 showing total budget increase and total percentage increase

Human service Department	Budget for 1975	Budget for 1978	Total dollar increase in budget	Total percent increase in budget
1	$1,000,000	$1,200,000	$200,000	20%
2	500,000	650,000	150,000	30%

the figures and the assumptions underlying the choice. Department 1 has
stayed relatively closer to its original 1975 budget than department 2 ("relatively" indicates that the percentage increase column should be used as
the basis of comparison). Thus, it could be said that department 1 has
done a better job than department 2. However, if the question were
shifted to which department costs less to support, then Department 2 may
be said to be doing a better job ("absolute costs" indicate that the dollar
increase column should be used as the basis of comparison). Because each
perspective is empirically correct yet leads to an opposite conclusion from
the other, both the percentage increase and the absolute cost figures
should be included in a report for further evaluative judgment.

A second problem with percentages is sample size. Twenty percent of
a sample of 1,000 is 200; 20 percent of a sample of 5 is 1. Both percentages
are calculated correctly, but a change of one subject in the sample of 1,000
is only one tenth of 1 percent, or .1 percent, whereas a change of one subject in the sample of 5 is a 20 percent change. Because single subjects have
less overall effect on percentages in larger samples, more confidence may
be placed on the conclusions drawn from them than from smaller samples. As a rule of thumb, one should not use percentages for sample sizes
of less than 30, or preferably 50.

A third problem with percentages is the representativeness of the
sample on which the percentages are based. If 59 percent of respondents
in a sample say that welfare people are avoiding work, this result is limited
by how the sample was drawn (was it a random sample to ensure a representative sample of the population or was it a biased sample?) and by
what population was sampled (who is represented? businessmen, divor-
cees, liberals, people over 30, welfare recipients, or the blind?). Much care
and restraint is required, then, before drawing generalizations from any
percentage of percentage distribution.

In sum, the purpose of percentages, the sample size used, and the ap-
propriateness of generalizations drawn from the sample must always be
made explicit before percentages may be accepted as meaningful and appropriate. This statement is also true of all statistics. Unless correct procedures are followed, it is all too easy to "lie with statistics." Many logical

steps and explicit assumptions are required to use statistics properly. If these are violated or ignored, it is still possible to calculate statistics, but it is not possible to interpret them correctly. A major goal of this book is to define the logical steps and assumptions necessary to use statistics competently and truthfully.

5.6 RATIOS

A *ratio* is a number (the numerator) divided by another number (the denominator).

Proportions and percentages are special types of ratios. A *proportion* is the frequency of a score interval (the numerator) divided by the total number of subjects (the denominator). A *percentage* is the frequency of a score interval (the numerator) divided by the total number of subjects (the denominator), the result of which is multiplied by 100.

Other types of ratios also exist. A *sex ratio,* for example, is the number of men in a group (the numerator) divided by the number of women in the same group (the denominator), the result of which is usually multiplied by 100 to get rid of two decimal spaces. A sex ratio of under 100 indicates that there are more women than men, a ratio of 100 indicates an equal number of men and women, and a ratio over 100 indicates that there are more men than women. The concept of a sex ratio may be extended to any other comparison of groups, as long as the number in one group (the numerator) is divided by the number in the other group (the denominator).

Rates are another type of ratio. A *rate* is the frequency of an occurrence divided by the potential number of occurrences, the result of which is multiplied by some multiple of 100 so that no decimal spaces will be needed. A birth rate, for example, is usually calculated by taking the number of live births in a year (the numerator) and dividing by the number of women of child-bearing age in that year (the denominator). Accident rates, recidivism rates, and admissions rates are other examples. It should be noted that rates are often reported as so many per 1,000 or 100,000, which means that the rate has been multiplied by 1,000 or 100,000 to get rid of three to five decimal places.

Lastly, rates of change are yet another type of ratio. A *rate of change,* as mentioned earlier, is the actual change observed during a specific time period (the numerator) divided by the original amount at the beginning of that time period (the denominator). An example of a positive rate of change was given in Section 5.5 while discussing Table 5.9. But a negative rate of change may occur as well. If the population of mental hospitals in a state was 80,000 in 1965 and 45,000 in 1975, the negative rate change would be:

$$\frac{\text{actual change}}{\text{original amount}} = \frac{\text{amount at time } 2 - \text{amount at time } 1}{\text{amount at time } 1}$$

$$= \frac{45,000 - 80,000}{80,000}$$

$$= -.44 \text{ or a } 44 \text{ percent decrease.}$$

Rates of change are often reported as percentage changes, which indicate that the calculated rate of change has been multiplied by 100 to get rid of two decimal spaces. Thus, the rate of change reported as $-.44$ would become a percentage change of negative 44 percent or a 44 percent decrease.

5.7 CUMULATIVE FREQUENCIES AND PERCENTAGES

A *cumulative frequency* (cf) is the sum of frequencies within a specified score interval and all frequencies in score intervals below it in actual score value. A *cumulative percentage* (cpc) is the sum of all percentages for a specified score interval *and* all score intervals below it in actual score value. The definition of cumulative frequencies and percentages thus implies that the score intervals of a variable are rank-ordered. Unlike frequencies and percentages, then, cumulative frequencies and cumulative percentages may only be used for ordinal or interval variables and not for nominal variables.

If cumulative frequencies are calculated for each score interval in a distribution, the result is called a *cumulative frequency distribution*. A cumulative frequency distribution is shown in column 5 of Table 5.10. The cumulative frequency for 10-year-old girls is 2; the cumulative frequency of girls 11 years old or younger is 5 (three 11-year-old girls plus two 10-year-old girls); the cumulative frequency for girls 12 years old or younger is 9 (four 12-year-old girls plus three 11-year-old girls plus two 10-year-old girls); and so on. Notice that the last cumulative frequency is always the number of people in the group; in this case, the group size is 30 and the cumulative frequency for girls in this home 17 years old or younger is thus 30 (two 17-year-old girls plus all the rest of the girls).

If cumulative percentages are calculated for each score interval, the result is called a *cumulative percentage distribution*. A cumulative percentage distribution is shown in column 6 of Table 5.10.

5.8 PERCENTILE RANKS

A *percentile rank* (pr) is a number indicating the percentage of scores in a distribution of rank-ordered score intervals that fall below the midpoint of a specified score interval. A percentile rank may only be used for ordinal or interval variables and not for nominal variables.

Table 5.10. Frequencies, proportions, percentages, cumulative frequencies, cumulative percentages, and percentile ranks of age for the girls in the detention home ($n = 30$)

Age	f	p	pc	cf	cpc	pr
10	2	.07	7	2	7	3
11	3	.10	10	5	17	12
12	4	.13	13	9	30	23
13	6	.20	20	15	50	40
14	6	.20	20	21	70	60
15	4	.13	13	25	83	77
16	3	.10	10	28	93	88
17	2	.07	7	30	100	97

A percentile rank is calculated as follows:

$$pr = \frac{(cf - \frac{f}{2})}{n} \times (100)$$

where

pr = percentile rank

cf = cumulative frequency for a specified score interval

f = frequency for the same specified score interval

$(cf - \frac{f}{2})$ = cumulative frequency from the *midpoint* of the specified score interval and all score intervals below it in value

n = total group size

$\frac{(cf - \frac{f}{2})}{n}$ = proportion of frequencies below the midpoint of the specified score interval, which is to be multiplied by 100.

Column 7 of Table 5.10 shows the calculated ranks for each age interval of the detention home girls. The percentile rank of 60, for example, indicates that 60 percent of the girls in the home are below age 14:

$$pr = \frac{(21 - {}^6/2)}{30} \times 100$$

$$= \frac{18}{30} \times 100$$

$$= .60 \times 100$$

$$= 60 \text{ percent.}$$

The definition and calculation of a percentile rank use the notion of true class limits as well as the assumption of scores being equally distributed within a score interval. The true class limits for age 14 are conceptualized as ranging from 13.5 to 14.5.[1] The six girls of this age are also assumed to be equally spread out within this score interval, so dividing by two presumably divides the number of subjects in the score interval in half. A percentile rank includes only those below the age midpoint of 14. A cumulative percentage, in contrast, includes all those below the age point of 14.5 or the upper class limit. The percentile rank of a score interval will thus be smaller than the cumulative percentage.

A comparison of columns 6 and 7 in Table 5.10 demonstrates that percentile ranks are smaller than cumulative percentages at each score interval. To illustrate this difference in another way, a percentile rank may be calculated from a cumulative percentage as follows:

$$pr = cpc - \frac{pc}{2}$$

where cpc = cumulative percentage for a specified score interval
           pc = percentage for the same score interval.

The percentile rank for girls age 14 is, thus, the same as before:

$$pr = 70 - \frac{20}{2}$$
$$= 70 - 10$$
$$= 60 \text{ percent.}$$

Percentile ranks may be used for many applied purposes. Scores from psychological tests are often reported this way. The Graduate Record Exam or Miller Analogies Test, for example, indicate the percentage of individuals who have scores less than a certain specified score. Percentile ranks may also be used to calculate various statistical measures such as the median (a measure of central tendency) and quartile range (a measure of variability).

5.9 SPSS

Many of the statistics in this chapter are calculated by the SPSS procedure card called FREQUENCIES. Read Chapters 7 and 8 in the *SPSS Primer*.

SPSS *procedure cards* call forth the desired statistical procedures. By specifying FREQUENCIES in the control field of a procedure card, the computer will provide univariate distributions for frequencies (called

---

[1]Using the convention of "age" meaning "age at last birthday," the true class limits for age 14 would be 14 to 15 rather than 13.5 to 14.5. However, the examples in this book consider age in the same way as most interval-level variables, so that the class limits for age 14 are treated as 13.5 to 14.5

"absolute freq" or "adjusted freq" on the printout), and cumulative percentages (called "cum freq" on the printout).

I used the following set of SPSS cards to get the univariate distributions for all variables in the detention home study:

1: Control field	16: Specification field
RUN NAME	DETENTION HOME STUDY
COMMENT	THIS SPSS COMPUTER PROGRAM WILL PROVIDE UNIVARIATE DISTRIBUTIONS FOR ALL VARIABLES IN THE DETENTION HOME STUDY
DATA LIST	FIXED (1)/1 SUBJECT 1-2, AGE 3-4, RELIGION 5, DAYS 6-7, OFFENSE 8
INPUT MEDIUM	CARD
N OF CASES	30
VALUE LABELS	RELIGION (1) PROTESTANT (2) CATHOLIC (3) OTHER/ OFFENSE (1) LOW (2) MODERATE (3) HIGH/
VAR LABELS	SUBJECTS, SUBJECT IDENTIFICATION NUMBER/ AGE, AGE AT TIME OF DETENTION/ RELIGION, RELIGION/ DAYS, NUMBER OF DAYS SPENT/ OFFENSE, SERIOUSNESS OF OFFENSE/
FREQUENCIES	GENERAL = AGE, RELIGION, DAYS, OFFENSE
OPTIONS	8
STATISTICS	ALL
READ INPUT DATA	
[place data deck here]	
FINISH	

The RUN NAME card tells the computer the study's title, which will be printed on the top of every page of the printout. The title may not be longer than 64 characters in length, including spaces.

The COMMENT card provides a note composed by the programmer to indicate what this program is about or simply to convey a message. It serves no further informational or computational purpose. It is an optional but useful card which may be inserted in the program at any place and as many times as desired. Read Section 12.2 in the *SPSS Primer*.

The DATA LIST through VAR LABELS cards have been discussed already in Section 4.5.

The FREQUENCIES card tells the computer which variables should be put into univariate distributions. The entry "GENERAL = " should be put at the beginning of the specification field and before the list of variables you want analyzed. Note that the subject identification number has not been put on this list of variables, because it only provides identification of each subject and has no other relevance to detention homes or girls in them.

The OPTIONS card is an optional card which allows choice about how the calculations should be done. To use an analogy, the FREQUEN-CIES card is a steak, and the OPTIONS card tells how it should be cooked: rare, medium, or well done. In this SPSS program, I chose to ask for FREQUENCIES with histograms (to be discussed in Chapter 11) as option 8.

The STATISTICS card is another optional card that tells the computer also to calculate measures of central tendency and measures of variability (to be discussed in Chapters 7 and 8). I chose to have all available statistics calculated, so I put ALL in the specification field of the STA-TISTICS card.

The READ INPUT DATA card tells the computer to read the computer cards that contain the data, as discussed in Section 4.4 of this book. Whenever data cards are used, the READ INPUT DATA card must be placed after the first set of procedure cards. In this case, the READ IN-PUT DATA card is placed after the STATISTICS card of the FRE-QUENCIES procedure.

The 30-card data deck should now be placed directly after the READ INPUT DATA card.

The FINISH card tells the computer that the end of the SPSS program has been reached. The FINISH card should always be the last card of the whole deck.

Tables 5.11 through 5.14 show how SPSS prints univariate frequency and percentage distributions for age, religion, number of days spent in the detention home, and seriousness of offense.

For each SPSS table, note that: 1) the RUN NAME title, DETEN-TION HOME STUDY, is printed at the top; 2) the name of the variable as identified on the FREQUENCIES card is printed next, along with its full name from the VAR LABELS card; 3) the score intervals are printed under the column called "CODE," along with any score names identified on the VALUE LABELS cards; 4) the frequencies for every score interval are printed in the column called "ABSOLUTE FREQ;" 5) the percentages for every score interval are printed in the columns "RELATIVE FREQ" and "ADJUSTED FREQ" (the ADJUSTED FREQ column ignores missing data in calculating percentages; and 6) the cumulative frequencies for every score interval are printed in the column called "CUM FREQ."

Statistics are printed underneath the univariate distributions. These statistics are measures of central tendency and measures of variability, which will be discussed in Chapters 7 and 8.

Table 5.11.   SPSS univariate distribution for age at time of detention

DETENTION HOME STUDY

AGE   AGE AT TIME OF DETENTION

CATEGORY LABEL	CODE	ABSOLUTE FREQ	RELATIVE FREQ (%)	ADJUSTED FREQ (%)	CUM FREQ (%)
	10.	2	6.7	6.7	6.7
	11.	3	10.0	10.0	16.7
	12.	4	13.3	13.3	30.0
	13.	6	20.0	20.0	50.0
	14.	6	20.0	20.0	70.0
	15.	4	13.3	13.3	83.3
	16.	3	10.0	10.0	93.3
	17.	2	6.7	6.7	100.0
	TOTAL	30	100.0	100.0	

MEAN	13.500	STD ERR	0.352	MEDIAN	13.500
MODE	13.000	STD DEV	1.925	VARIANCE	3.707
KURTOSIS	− 0.633	SKEWNESS	0.000	RANGE	7.000
MINIMUM	10.000	MAXIMUM	17.000		

VALID CASES   30   MISSING CASES   0

Table 5.12.   SPSS univariate distribution for religion

DETENTION HOME STUDY

RELIGION   RELIGION

CATEGORY LABEL	CODE	ABSOLUTE FREQ	RELATIVE FREQ (%)	ADJUSTED FREQ (%)	CUM FREQ (%)
PROTESTANT	1.	15	50.0	50.0	50.0
CATHOLIC	2.	11	36.7	36.7	86.7
OTHER	3.	4	13.3	13.3	100.0
	TOTAL	30	100.0	100.0	

MEAN	1.633	STD ERR	0.131	MEDIAN	1.500
MODE	1.000	STD DEV	0.718	VARIANCE	0.516
KURTOSIS	− 0.699	SKEWNESS	0.692	RANGE	2.000
MINIMUM	1.000	MAXIMUM	3.000		

VALID CASES   30   MISSING CASES   0

Table 5.13.   SPSS univariate distribution for days spent

DETENTION HOME STUDY

DAYS   NUMBER OF DAYS SPENT

CATEGORY LABEL	CODE	ABSOLUTE FREQ	RELATIVE FREQ (%)	ADJUSTED FREQ (%)	CUM FREQ (%)
	1.	1	3.3	3.3	3.3
	2.	1	3.3	3.3	6.7
	3.	1	3.3	3.3	10.0
	4.	4	13.3	13.3	23.3
	5.	4	13.3	13.3	36.7
	6.	2	6.7	6.7	43.3
	7.	3	10.0	10.0	53.3
	8.	3	10.0	10.0	63.3
	9.	5	16.7	16.7	80.0
	10.	1	3.3	3.3	83.3
	11.	2	6.7	6.7	90.0
	13.	1	3.3	3.3	93.3
	13.	1	3.3	3.3	96.7
	14.	1	3.3	3.3	100.0
	TOTAL	30	100.0	100.0	

MEAN	7.167	STD ERR	0.591	MEDIAN	7.167
MODE	9.000	STD DEV	3.239	VARIANCE	10.489
KURTOSIS	− 0.487	SKEWNESS	0.199	RANGE	13.000
MINIMUM	1.000	MAXIMUM	14.000		

VALID CASES   30   MISSING CASES   0

Table 5.14.   SPSS univariate distribution for seriousness of offense

DETENTION HOME STUDY

OFFENSE   SERIOUSNESS OF OFFENSE

CATEGORY LABEL	CODE	ABSOLUTE FREQ	RELATIVE FREQ (%)	ADJUSTED FREQ (%)	CUM FREQ (%)
LOW	1.	12	40.0	40.0	40.0
MODERATE	2.	9	30.0	30.0	70.0
HIGH	3.	9	30.0	30.0	100.0
	TOTAL	30	100.0	100.0	

MEAN	1.900	STD ERR	0.154	MEDIAN	1.833
MODE	1.000	STD DEV	0.845	VARIANCE	0.714
KURTOSIS	− 1.585	SKEWNESS	0.198	RANGE	2.00
MINIMUM	1.000	MAXIMUM	3.000		

VALID CASES   30   MISSING CASES   0

Chapter 6

Graphs of Univariate Frequency Distributions

Graphs are techniques for describing data by pictures and are useful for two reasons. First, most people can read and understand graphs with little or no knowledge of statistics. Second, because graphs visually display data, the information they convey is easier to remember— one picture is worth a thousand words or numbers. Both reasons suggest that graphs should be used to summarize important points about your data and to present information to general audiences.

There are a number of different types of graphs. Before making a graph, however, either a univariate frequency or a percentage distribution is necessary to provide information on the size of the variables' score intervals and the frequency or percentage with which each score interval occurs. Given this distribution, two more decisions need to be made before actually making a graph. First, a decision must be made on what variables should be put into graph form. Because graphs take a lot of space and too many graphs may be overwhelming to an audience, graphs should only be used to highlight the most important variables. This decision is the same as that of a good storyteller, who must make the crucial points of the story stand out from all the surrounding descriptive narrative. Second, a decision must be made on which graph is appropriate for the level of measurement of the variable. Figure 6.1 summarizes this last decision making step. For nominal variables, either pie charts or bar graphs are appropriate. For ordinal and interval variables, histograms, line graphs, frequency polygons, and curves are most appropriate, although pie charts and bar graphs may also be used. Histograms and line graphs make it easier to read and compare the relative frequency of score intervals, whereas frequency polygons and curves highlight the overall shape of a distribution.

No matter which graph is chosen, however, every graph is based on one common principle. This principle is that all the space or area contained within the drawn graph sums to 1.00 (as proportions do) and represents the total size or 100 percent of the sample. The area within the graph is then divided into sections, which should be in proportion to the fre-

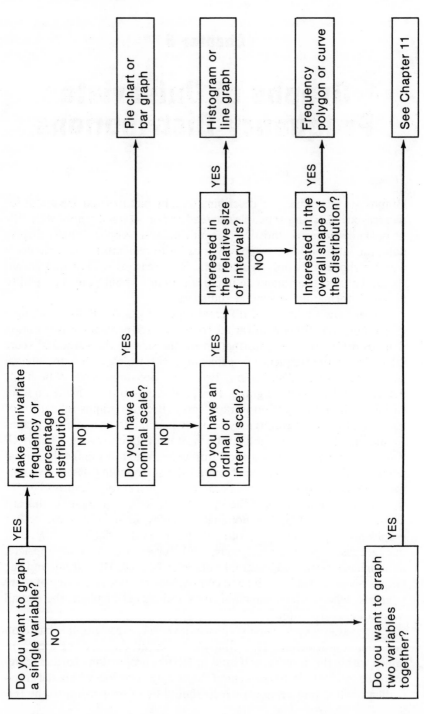

Figure 6.1.   Decision-making chart for graphs of univariate distributions.

quency of score intervals. The sum of these sections is 100 percent of the graph area, just as the sum of frequencies is 100 percent of the sample. The rest of this chapter discusses how to draw each type of graph.

6.1 PIE CHARTS

A *pie chart* uses a circle to represent the whole sample. Each wedge or piece of the pie represents a score interval of the variable being graphed. The size of each wedge represents the relative size or percentage of subjects in the score interval. The sum of percentages for all score intervals is 100 percent, the total area of the pie circle.

Several steps are necessary to draw a pie chart. In describing each step, the univariate frequency distribution of religion for the 30 girls in the detention home will be used as an example. Note that religion is a nominal variable.

The first step in making a pie chart is to calculate the proportion or percentage of subjects in each score interval or wedge, as in Table 6.1. Fifty percent of the girls are Protestant, 37 percent are Catholic, and 13 percent are of some other religion. These percentages may also be expressed as proportions as shown in column 3 of Table 6.1.

The second step is to calculate the angle size of each score interval or wedge. The total angle of a whole circle is 360 degrees (360°). The total angle must be divided into smaller angles that represent the proportion of each score interval. This is done by multiplying the proportion of scores in each score interval by 360 degrees, as in the second column of Table 6.2. The angle size for Protestant girls is 180 degrees, the angle size for Catholic girls is 133 degrees, and the angle size for girls of other religions is 47 degrees.

The last step is to use these angle sizes to divide the circle into three wedges, with each wedge labeled by the name of the score interval it represents and the percentage of girls in that interval. This has been done in Figure 6.2. Because each wedge is drawn in proportion to the frequency of subjects in the score interval, it is easy to tell at a glance that Protestant girls form the largest group in the detention home, Catholic girls are the next largest group, and girls of other religions form the smallest group.

Table 6.1. Percentage distribution of religion for the girls in the detention home ($n = 30$)

Religion	Percentage	Proportion
Protestant	50% (15)[a]	.50
Catholic	37% (11)	.37
Other	13% (4)	.13

[a] Frequencies are in parentheses.

Table 6.2.   Calculation of the wedge angle size in a pie chart for religion of girls in the detention home ($n = 30$)

Religion	Calculation of angle size (proportion × 360 degrees)	Angle size of wedge
Protestant	.50 × 360° =	180°
Catholic	.37 × 360° =	133°
Other	.13 × 360° =	47°

In sum, although pie charts may be used with any type of variable (nominal, ordinal, and interval), they are used primarily for nominal variables.

6.2   COORDINATE AXES

Many of the following graphing techniques use *coordinate axes* to graph data. Coordinate axes are two straight lines placed at right angles to each other. The horizontal line is called the *X axis* or *abscissa*. The vertical line is called the *Y axis* or *ordinate*. The point at which the *X* and *Y* axes cross or intersect each other is called the *origin*. The origin is the zero or starting point for each axis. Figure 6.3 shows what coordinate axes look like.

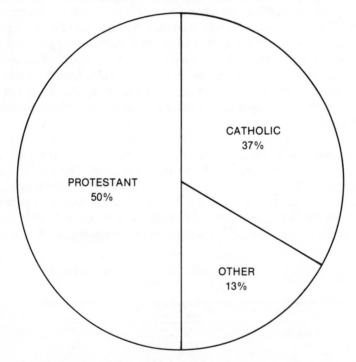

Figure 6.2.   Pie chart for religion of girls in the detention home ($n = 30$).

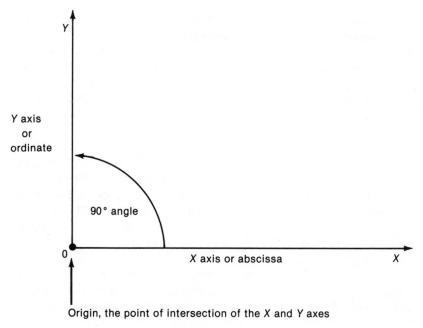

Figure 6.3.   An example of coordinate axes.

For graphs of univariate distributions, the horizontal or X axis represents the score intervals of the variable. These score intervals should be evenly spaced along the axis. Since the origin always represents the value of zero, the lowest valued score intervals of ordinal and interval variables should be placed at the left and move toward the right as the score intervals increase in value. The score intervals of nominal variables may be placed in any order. The vertical or Y axis represents the frequency or percentage observed for each score interval. The frequencies or percentages should be evenly spaced and increase in equal steps from the origin. Again, since the origin always represents the value of zero, the lowest frequency or percentage should be placed at the bottom and, as the values increase, move toward the top of the vertical line.

All the graphs that follow are based on coordinate axes.

6.3   BAR GRAPHS

A *bar graph* uses rectangles, called bars, within the coordinate axes to illustrate the frequency or relative size of each score interval.

Several steps are required to draw a bar graph. The first step is to calculate the frequency and percentage of subjects within score intervals as in the distribution for religion in Table 6.1.

The second step is to draw coordinate axes. The X axis represents the score intervals for the variable. For religion, the score intervals are Protestant, Catholic, and other as in Figure 6.4. Since religion is a nominal variable, the score intervals may be arranged in any way. The Y axis represents the frequency or number of subjects observed in each score interval. Because the Y axis need not be labeled much higher than the number of subjects in the most frequently observed score interval, the Y axis in Figure 6.4 has been labeled zero at the origin and then goes up to only 20 subjects in equal steps of 5.

The third step is to draw rectangular bars within the coordinate axes. Figure 6.4 shows how this should be done. Bars are drawn for each score interval by: 1) placing the middle of each bar over the presumed middle of the score interval, which is marked by a slash on the Y axis; 2) making the height or length of each bar along the Y axis equivalent to the frequency of the score interval; 3) making the width of each bar the same; and 4) leaving a space between each bar.

The last step is to label the bar graph. The title of the graph should appear at the top and include the type of graph (a bar graph), the variable being graphed (religion), the subjects (girls in the detention home), and

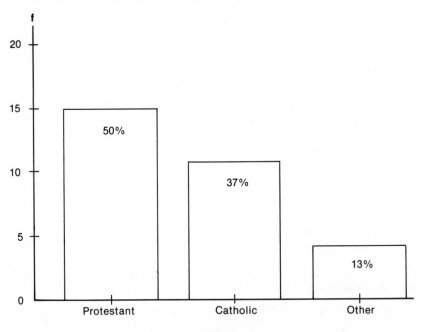

Figure 6.4.   A bar graph of religion for girls in the detention home ($n = 30$).

the number of subjects ($n = 30$) as in Figure 6.4. The X axis should have equally spaced marks along it to represent score intervals. The name of each score interval should be placed directly below the marks. If the first score interval used is much greater than zero and subsequent score intervals proceed at the same size, the X axis may be broken by a double slash ($//$) near the origin to indicate that the numbering will begin at a point after which the size of intervals will be the same. The name of the variable being graphed should appear further below the X axis. The Y axis should have equally spaced marks along it to represent frequencies or percentages. The frequency or percentage associated with each mark should be placed next to the mark. Not every frequency or percentage need be put in, however. Rather, only a sufficient number of slashes and labels should be drawn to allow the reader to count frequencies or percentages accurately. It is assumed that frequency or percentage labels will begin at zero and increase in regular intervals. However, if every score interval has a large number of subjects in it, the Y axis may be broken by a double slash ($//$) near the origin to indicate that the numbering will begin at 30, 60, or some other number and thereafter proceed in regular intervals. The label f for frequency, or % for percentage, should also be placed at the top of the Y axis to indicate which counting procedure has been used. Lastly, the percentage of subjects in each score interval should be written inside the top of each bar so the reader can compare the relative size of score intervals visually and in terms of percentages.

In sum, like pie charts, bar graphs may be used for any type of variable. They are used primarily for nominal variables, however. Bar graphs also tend to be used more than pie charts because they are easier to draw and because more score intervals may be included without making the graph difficult to read.

6.4   HISTOGRAMS

A *histogram* is a bar graph with no spaces between bars. The lack of spaces between bars represents the relative continuity between score intervals for both ordinal and interval scales. Otherwise, a histogram is drawn just as a bar graph is.

Figure 6.5 shows a histogram for the interval variable called "age" in the detention home study. (This histogram is based on the age distribution in Table 5.1.) Note that each bar is centered at the midpoint of each score interval, the left side of each bar is at the lower true class limit of the interval, and the right side of each bar is at its upper true class limit. The bar for age 15, for example, is centered at the midpoint 15. The left side of the bar is at 14.5 and the right side of the bar is at 15.5. Note also that a double slash appears at the beginning of the X axis. This double slash

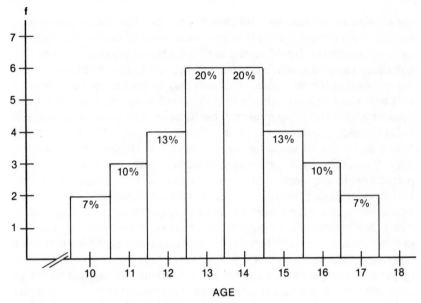

Figure 6.5.   A histogram of age for girls in the detention home ($n = 30$).

indicates that no girls were below the age of 10 and thus, to save space, ages 1 through 9 are not shown on the graph.

6.5   LINE GRAPHS

A *line graph* uses vertical lines rather than bars to show the frequency or percentage of score intervals. Vertical lines are placed at the midpoint of each score interval, and their height represents the frequency or percentage of each interval. Figure 6.6 is an example of a line graph. Line graphs are used primarily when a variable has many score intervals, whereas histograms are used when there are fewer or grouped score intervals. Choice between them, however, is actually based on user preference.

6.6   FREQUENCY POLYGONS

A *frequency polygon* is used to highlight the overall shape of a distribution. It is called a polygon because it has many connected straight sides.

Drawing a frequency polygon is much like drawing a line graph. Dots are placed above the midpoint of each score interval, and their height represents the frequency or percentage of each interval. The dots are then connected by straight lines. The first and last dots should also be connected to the X axis, one score interval below and above the score inter-

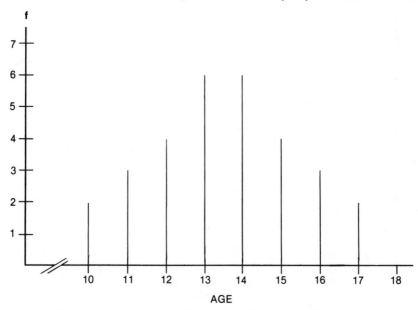

Figure 6.6.   A line graph of age for girls in the detention home ($n = 30$).

vals used, by straight lines. Figure 6.7 is an example of a frequency
polygon for age distribution.

6.7   CURVES

Curves, like frequency polygons, are used to highlight the overall shape of
a distribution. Unlike frequency polygons, however, curves eliminate
angular roughness by drawing a freehand, smooth curve which connects
the dots only approximately. This procedure forces the eye to concentrate
on the general rather than specific shape of a distribution. Figure 6.8 is an
example of a curve for age distribution.

6.8   DISTRIBUTION SHAPES

Concentration on the general shape of curves suggests several broad
classifications of *distribution shapes*. First, distributions may be classified as symmetrical or skewed. A distribution is said to be *symmetric*
when the two halves are exactly alike. If you fold a curve along its center
line, both halves should correspond totally—much like cutting out a
snowflake with scissors. Distribution A in Figure 6.9 is a symmetric curve.
A distribution is said to be *skewed* when the two halves or sides are not
alike. If few frequencies are on the left side and many more are on the

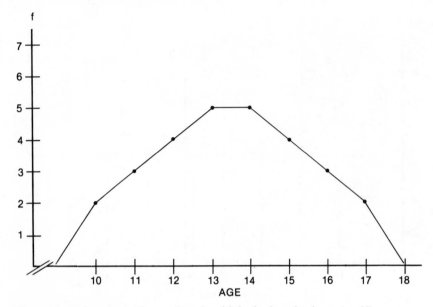

Figure 6.7.   A frequency polygon of age for girls in the detention home (*n* = 30).

right side, the distribution is called *negatively skewed* or *skewed to the left* (see distribution B). If few frequencies are on the right side and many more are on the left side, the distribution is called *positively skewed* or *skewed to the right* (see distribution C).

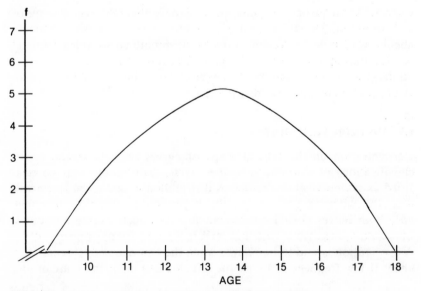

Figure 6.8.   A curve of age for girls in the detention home (*n* = 30).

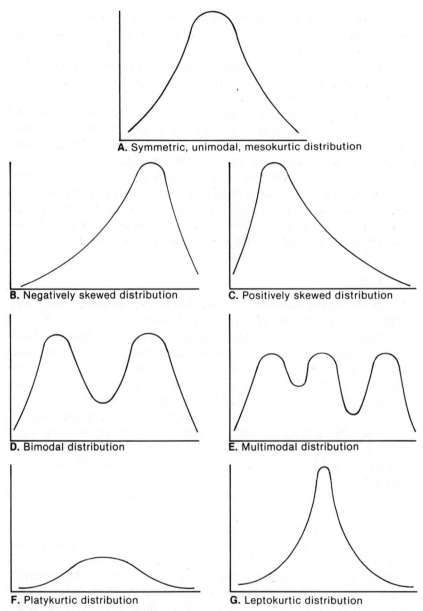

A. Symmetric, unimodal, mesokurtic distribution

B. Negatively skewed distribution

C. Positively skewed distribution

D. Bimodal distribution

E. Multimodal distribution

F. Platykurtic distribution

G. Leptokurtic distribution

Figure 6.9.   Types of distributions.

Second, a distribution may be classified in terms of its number of peaks or *modes*. A distribution with one peak, as in distribution A, is called *unimodal*. A distribution with two peaks, as in distribution D, is called *bimodal*. A distribution with three peaks or more, as in distribution E, is called *multimodal*.

Third, distributions may be classified in terms of the degrees of their peakedness or *kurtosis*. A flat curve or a curve with a relatively small degree of peakedness, as in distribution F, is called *platykurtic*. A distribution with a moderate degree of peakedness, as in distribution A, is called *mesokurtic*. A distribution with a high degree of peakedness, as in distribution G, is called *leptokurtic*.

The usefulness of various statistics, descriptive and inferential, will vary according to the type of distribution shape your population has. Although the problems associated with each distribution in descriptive statistics will be discussed in the following chapters, two general statements may be made here: 1) a symmetrical, unimodal, mesokurtic distribution is called a *normal distribution* and is of central importance in statistics; and 2) a distribution that is skewed or multimodal severely limits the use of many descriptive statistics. The degree of kurtosis has little effect on the choice of descriptive statistics to be used.

6.9   SPSS

One of the graphs that SPSS produces is a histogram in the FREQUEN-CIES program (see page 67 in the *SPSS Primer*).

Strictly speaking, the SPSS histogram is a line graph, because a line is used instead of a bar. Another difference is that SPSS prints the score intervals (codes) on the *Y* axis and the frequencies on the *X* axis. This has been done for the sake of convenience in printing the computer analysis. When SPSS histograms are used in a report, however, they should be redrawn to follow the terms and conventions discussed in this chapter.

The SPSS FREQUENCIES program for the detention home study given in Section 5.9 will provide SPSS histograms for every variable. The SPSS card for histograms is the OPTIONS card with selection of option 8.

Table 6.3 shows an example of a SPSS histogram for the ordinal variable "seriousness of offense."

Use of the STATISTICS card in the FREQUENCIES program also provides the degree of skewness and kurtosis in a distribution. The measure of skewness in SPSS results in zero if a distribution is symmetric and bell-shaped, a negative value if negatively skewed, and a positive value if positively skewed. The measure of kurtosis in SPSS results in zero if a distribution is mesokurtic, a negative value if platykurtic, and a positive value if leptokurtic. Tables 5.11 through 5.14 give these values for all the distributions in the detention home study. None of these distributions is so overly skewed or kurtotic to be a problem. However, it should be apparent that these values are meaningless for religion, which is a nominal variable. It should be noted that there is no measure to determine unimodality other than the "eyeball" method.

Table 6.3.   SPSS histogram for seriousness of offense

OFFENSE   SERIOUSNESS OF OFFENSE

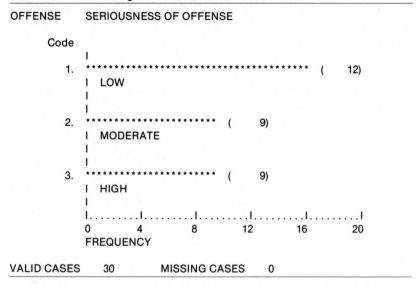

```
    Code
        I
     1. ***************************************** (    12)
        I  LOW
        I
        I
     2. ********************* (     9)
        I  MODERATE
        I
        I
     3. ********************* (     9)
        I  HIGH
        I
        I........I........I........I........I........I
        0        4        8       12       16       20
        FREQUENCY
```

VALID CASES   30   MISSING CASES   0

Chapter 7

Measures of
Central Tendency

Four sources of information may be used to describe univariate frequency distributions:

1. the univariate frequency distribution itself
2. the graphic representation of its shape
3. the measure of its central tendency or "typical" score
4. the measure of its variability or spread of scores.

The first source of knowledge, the univariate frequency distribution itself, is obviously the most complete. However, by its very completeness it is awkward to use, just as it would be awkward to use a picture of every tree in a forest to describe a forest. Description is made less awkward by the use of graphs, measures of central tendency, and measures of variability.

The most important aspect of graphs is the shape of a distribution, particularly its degree of skewness and its number of peaks or modes. The most important measures of central tendency are the mode, the median, and the mean. The most important measures of variability are the variation ratio, range, variance, and standard deviation.

Whereas the previous chapter discussed graphs in reference to the shape of distributions (see Section 6.8 especially), this chapter focuses on measures of central tendency. The following chapter concentrates on measures of variability. Together, Chapters 6, 7, and 8 form one cornerstone of descriptive statistics. These methods not only make it easier to describe a distribution, they also make it possible to compare two or more distributions. They are, therefore, basic to the understanding and use of statistics.

7.1 DEFINITION

Measures of central tendency are single numbers that describe the typical score in a distribution. These single numbers are points on the same scale of measurement as the variables. Their purpose is to provide a summary measure of the most typical score (the central tendency) in a distribution, or, in other words, the central point around which scores cluster. Figure 7.1 shows two distributions differing in central tendency. Assuming the same scale of measurement along the X axis, measures of central tendency for distribution A should be smaller or lower than for distribution B.

There are several measures of central tendency, each of which describes the central or typical point differently. The *mean* is the average of all scores in a distribution. The *median* is the scale point that divides the distribution of subjects in half. The *mode* is the most frequently occurring score.

Choice among these measures depends primarily on the scale of measurement for the variable, on the shape of the distribution, and on the purpose of your research. Figure 7.2 summarizes these choices. If you have an interval variable, and the shape of its distribution is relatively symmetrical, then the mean is the appropriate choice. If the shape of its distribution is badly skewed or if you have an ordinal variable, then the median is the more appropriate choice. However, if inferential statistics (see Part III) are also going to be used, it is common practice to calculate the mean for ordinal scales, although there are no mathematical grounds for doing so. In this situation, calculation of the mean is usually done as a preliminary step to more complex statistics that assume interval level variables. More will be said about this later, but for now this decision is indicated in Figure 7.2 by a dotted line. If you have a nominal variable with three or more categories, the mode is the only appropriate choice. However, if an ordinal or a nominal variable has only two categories, either because it was scored that way originally or because its categories

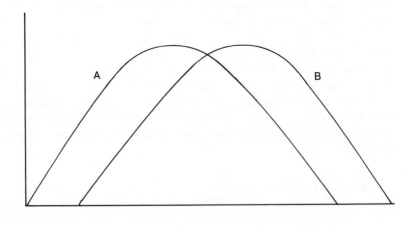

Figure 7.1.   Two distributions differing in central tendency.

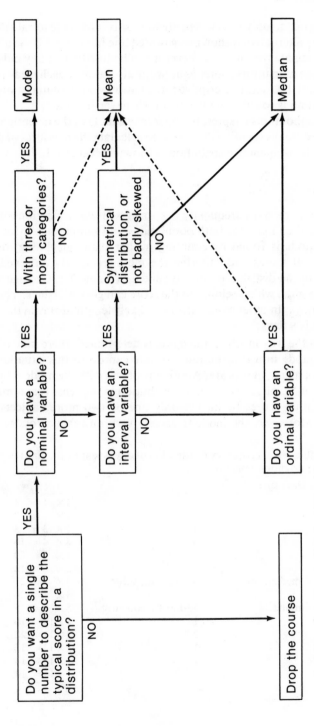

Figure 7.2.   Decision-making chart for measures of central tendency.

were collapsed to only two, it is mathematically legitimate to calculate the mean. This alternative is shown as a dotted line in Figure 7.2. As a rule of thumb, calculate the mean whenever it can be justified. If any indecision remains, use as many measures as appropriate, because each measure provides potentially useful descriptive information. In inferential statistics, a rule of thumb is to calculate the mean whenever justifiable.

Throughout this chapter, the interval variable in the detention home study called "days spent" will be used to calculate the measures of central tendency. Its frequency distribution is shown in Table 7.1.

7.2 MODE

The *mode* is the most frequently occurring score in a distribution. The symbol for the mode, for both populations and samples, is "Mo."

The mode is found by counting (rather than calculating) the frequency of all scores and choosing the most frequently observed score category. In the distribution of days spent in Table 7.1, the highest frequency is 5 girls, which belongs to the score category of 9 days. Thus, the mode is 9 days. In other words, the most likely length of stay in the detention home is 9 days.

Given the way in which the mode is determined, there may be more than one mode in a distribution. All one can do in this situation is to report what all the modes are. Another problem with the mode is that it is often a poor "typical" score of a distribution in the sense that it may differ from other scores by having only one or two more subjects in its category. Moreover, the mode is usually very unstable across samples,

Table 7.1. Frequency distribution of days spent by the girls in the detention home ($n = 30$)

Days spent	f	$f_j \times Y_j =$
1	1	$1 \times\ 1 =\ \ 1$
2	1	$1 \times\ 2 =\ \ 2$
3	1	$1 \times\ 3 =\ \ 3$
4	4	$4 \times\ 4 =\ 16$
5	4	$4 \times\ 5 =\ 20$
6	2	$2 \times\ 6 =\ 12$
7 ←median, mean	3 ←median, subject	$3 \times\ 7 =\ 21$
8	3	$3 \times\ 8 =\ 24$
9 ←mode	5 ←most frequent	$5 \times\ 9 =\ 45$
10	1	$1 \times 10 =\ 10$
11	2	$2 \times 11 =\ 22$
12	1	$1 \times 12 =\ 12$
13	1	$1 \times 13 =\ 13$
14	1	$1 \times 14 =\ 14$
	$n = 30$	$\Sigma f_j Y_j = 215$

because a change in only one subject's score may change the mode. Overall, then, these problems suggest that the mode is the least preferred measure of central tendency.

7.3 MEDIAN

The *median* is the scale point that divides the distribution in half such that there are an equal number of subjects above and below the median scale point. The symbol for the median, for both a population and a sample, is "Md."

The median is calculated from a frequency distribution, where the score intervals are ordered from low to high, by the following formula:

$$Md = L + (\frac{.5n - cf}{f})W$$

where:

L = the lower class limit of the median interval, which is found by locating the score interval where the median subject ($.5n$) is
$.5n$ = the median subject, or one-half of the population or sample size (n)
cf = the cumulative frequency of all score intervals below the median interval
f = frequency of the median interval
W = width of the median interval.

To find the median for days spent, the score intervals are first arranged from low to high as in Table 7.1. The median subject is $15 = .5n = .5(30)$. The median interval is 7 days, which is found by locating the score interval with the median subject (the 15th girl). The cumulative frequency of all score intervals below this median interval (cf) is $13 = 1 + 1 + 1 + 4 + 4 + 2$. The frequency of the median interval (f) is 3. Because 7 days represents the score interval with the true class limits of 6.5 and 7.5, the lower class limit of the median interval (L) is 6.5. It is assumed that the three subjects are equally spread out within this interval. As such, the calculation within the parentheses of the median formula gives the proportion of subjects (.667 of 3 girls) between the lower class limit of the median interval and the median point. The width of this median interval (W) is one (1.0) day, or the distance between 6.5 and 7.5. The width is multiplied by the proportion within the parentheses to give the actual distance ($.667 \times 1.0 = .667$ days) between the lower class limit and the median point. When this distance is added to the lower class limit of the median interval (L), you have the median point. Thus:

$$Md = 6.5 + (\frac{15 - 13}{3}) 1.0$$
$$= 6.5 + (.667) 1.0$$
$$= 7.167.$$

The median is 7.167 days. Fifty percent of the subjects in this distribution fall below and 50 percent fall above this median point. This point is the most representative score in the distribution in the sense of its being closest to every other score.

The above formula may be simplified when scores for all subjects in a data list are listed in order from low to high. When there is an odd number of subjects, the median is the score in the middle. When there is an even number of subjects, the median is the average of the two middle scores.

The median is used primarily for describing the central tendency of ordinal variables (especially if no further statistical analysis is to be done) and for interval variables when the distribution is markedly skewed.

7.4 MEAN

The *mean* is the arithmetic average of all the subjects' scores. The symbol for the mean of a population is the Greek letter mu (μ). The symbol for the mean of a sample is any capital letter with a bar over it, such as $\overline{Y}$. The capital letter Y represents a variable, and the bar over it indicates the mean of that variable.

If the mean is to be used in a description of a *population* of subjects, then it is calculated as:

$$\mu = \frac{\Sigma Y_i}{N}.$$

If the mean is to be used in a description of a *sample* of subjects, with the purpose of estimating the unknown population mean, then it is calculated as:

$$\overline{Y} = \frac{\Sigma Y_i}{n}$$

where:

Y_i = scores for all subjects, $i = 1$ to either N or n
ΣY_i = total sum of all scores in the distribution (see Sections 3.5–3.7)
N = population size
n = sample size.

Because the calculation of the mean does not differ for a population or for a sample, I will use the symbol for a sample mean unless a clear distinction is necessary.

The definition and formula for the mean indicate that the mean may be calculated most directly from a data list as in column 2 of Table 7.2. Add all subjects' scores for the variable "days spent" to get the total sum of scores ($\Sigma Y_i = 215$) and divide this sum by the total number of subjects ($n = 30$):

$$\overline{Y} = \frac{215}{30}$$
$$= 7.167 \text{ days.}$$

The mean number of days spent in the hypothetical detention home is 7.167 days. This mean or average is typical in the sense that it is the score every subject would have if the total number of days spent by everyone in the detention homes were divided equally for each girl.

If a frequency distribution rather than a data list is used to calculate the mean, the formula changes somewhat:

$$\overline{Y} = \frac{\Sigma f_j Y_j}{n}$$

where:

j = score intervals in the distribution, with j = first interval through the last interval
f_j = frequency of each score interval
Y_j = midpoint of each score interval
$\Sigma f_j Y_j$ = the total sum of every score interval's frequency multiplied by its score midpoint (see Sections 3.5–3.7).

The last column in Table 7.1 may be used to calculate the mean in this way:

$$\overline{Y} = \frac{215}{30}$$
$$= 7.167 \text{ days.}$$

Note that the calculation of the mean using a data list and frequency distribution result in the same answer. The same answer will only occur, however, if the score intervals in a frequency distribution are not grouped. Thus, for example, if the mean were calculated for the 1-week grouped intervals for days spent in Table 5.6, the result would be:

$$\overline{Y} = \frac{(16 \times 4) + (14 \times 11)}{30}$$
$$= 7.267 \text{ days.}$$

The reason for the discrepancy is that, for data lists and frequency distributions with ungrouped intervals, every subject's exact score is included in the calculations. For frequency distributions with grouped intervals, however, every subject's score is not known exactly and must therefore be estimated by the midpoint of the grouped interval. The larger the grouped interval, the greater the error. Whenever possible, then, means should be calculated from data lists or from frequency distributions with ungrouped intervals to give the most accurate result.

Note that the mean and median for days spent are both 7.167. The mean and median will be quite close for relatively symmetrical distributions. When the mean and median are the same value, the distribution is symmetrical. A symmetrical distribution is not necessarily unimodal, however. Whereas the distribution is unimodal for days spent, it might have been bimodal as in distribution A in Figure 7.3. When the mean, median, and mode are the same value, then the distribution is symmetrical and unimodal as in distribution B in Figure 7.3. When a distribution is skewed, as in distributions C and D, the mean, median, and mode will differ. The mode, as the most frequently occurring score, is least affected by extreme scores in the narrow tail of a skewed distribution. The median is affected, but not very much, because it is only the point at which 50 percent of the subjects fall above and below. The greater the *number* of extreme scores, however, the more they will pull the median towards the

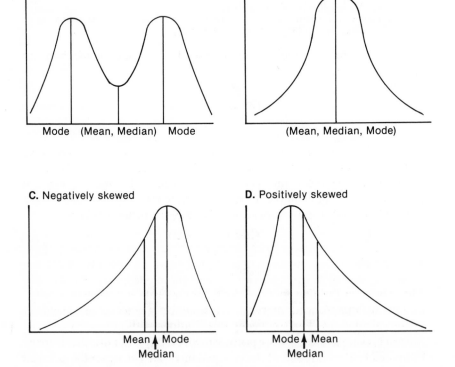

Figure 7.3.   The median, mean, and mode in four distributions.

tail. The mean is most affected by extreme scores because it is the arithmetic average of *all* scores. Thus, the mean will be pulled the furthest out toward the tail. In the case of a skewed distribution, then, it is helpful to report all three measures of central tendency.

Finally, it was mentioned at the beginning that the mean may be calculated legitimately for variables that have only two categories, be they nominal or ordinal. Any variable that has only two categories is called a *dichotomous* or *dummy* variable. A one is arbitrarily assigned to one of the categories and a zero to the other category. The mean is always the proportion of subjects in the category labeled with a 1. The calculation of the mean for a dummy variable is legitimate because no rules of measurement are violated (it classifies, it orders, and it has only one interval which must therefore be an equal interval) and because the resulting proportion is meaningful. For example, religion in the detention home study is a nominal variable with three categories. As such, it is only appropriate to use the mode as a measure of central tendency. If, however, the Protestant category were assigned a 1 and if the categories Catholic and other were collapsed into a category called "all others" and assigned a zero, the mean could be calculated. The numerator of the mean would be the sum of 15 ones for the 15 Protestant girls and 15 zeros for the 15 "all other" girls. Thus, the mean would be the proportion of Protestant girls:

$$\overline{Y} = \frac{15}{30}$$
$$= .50.$$

That is, 50 percent of the girls in this detention home are Protestant. Not only is this statement about the mean easily understood, but by using the mean rather than the mode, more complex data analysis is possible later on.

7.5 PROPERTIES OF THE MEAN

There are two important properties of the mean. Neither property is intuitively obvious, but both are essential to know.

First, the mean, as an arithmetic average, is so defined that *the sum of deviations about the mean is zero*. A *deviation* is the difference between a score (Y_i) and the mean of all scores (μ), or $(Y_i - \mu)$. A lowercase italic letter, such as y, is often used to represent a deviation score: $y = (Y_i - \mu)$. The sum of deviations is simply the total of all deviation scores added together. The sum of deviations is always equal to zero. In symbols, this first property of the mean looks like:

$$\Sigma(Y_i - \mu) = 0$$
or
$$\Sigma y = 0.$$

The second property of the mean, called the *least-squares* property, is that *the sum of squared deviations about the mean is a minimum value.* The sum of squared deviations tells you to take the deviation of every subject's score from the mean, square the deviation, and then add all squared deviations together. The result is a minimum value but not necessarily zero. If you took deviations around any other value than the mean, the result would be larger than this minimum value. In symbols, this secondary property of the mean looks like:

$$\Sigma(Y_i - \mu)^2 \text{ is a minimum deviation value}$$

or

$$\Sigma y^2 \text{ is a minimum deviation value}$$

Although the population mean (μ) was used in the definition of both properties of the mean, the sample mean ($\overline{Y}$) could be used instead.

Rather than prove either property of the mean, look at the arithmetic example for "days spent" in Table 7.2. The first column represents the identification numbers of the 30 subjects. The second column provides the score on days spent for each subject. The sum of scores (ΣY_i) for all 30 subjects is 215 days, which, when divided by the number of subjects, results in the mean ($\overline{Y}$) of 7.167 days spent. The third column shows the calculation of deviation scores in which the mean is subtracted from each subject's score on days spent. Given a little rounding error, addition of all the deviation scores, $\Sigma(Y_i - \overline{Y})$, in the third column results in zero just as expected under property 1 of the mean. Zero occurs because the mean acts as the center of gravity for the added scores such that the sum of negative deviations about the mean equals the sum of positive deviations about the mean. In the present example, the sum of negative deviations for scores below the mean is -39.7, and the sum of positive deviations for scores above the mean is $+39.7$. These, when added together, result in zero. The fourth column shows the calculation of squared deviation scores about the mean for each subject. The sum of squared deviations, $\Sigma(Y_i - \overline{Y})^2$, is 304.181. This number is referred to as the minimum value. The fifth and last column shows the calculation of squared deviation scores about the number of 1.00 rather than the mean of 7.167. The sum of these latter squared deviation scores is 1445.000; a sum that is larger than the minimum sum of squared deviations about the mean, or 304.181. This is as expected by property 2 of the mean.

7.6   USE

Knowing how to define and calculate measures of central tendency, the next question is: of what use are they from an applied perspective? Recall that the mean is the score every person would have if the total were di-

Table 7.2. A data list of number of days spent by girls in the detention home ($n = 30$)

Subject identification	Days spent (Y_i)	Deviations ($Y_i - \bar{Y}$)	Deviation score (y)	Squared deviations ($Y_i - \bar{Y}$)² = y^2	Squared deviations around 1 ($Y_i - 1$)²
1	5	$5 - 7.167 =$ -2.167	-2.167	$-2.167^2 =$ 4.696	$(5-1)^2 =$ 16
2	11	$11 - 7.167 =$ 3.833	3.833	$3.833^2 =$ 14.692	$(11-1)^2 =$ 100
3	8	$8 - 7.167 =$ $.833$	$.833$	$.833^2 =$ $.694$	$(8-1)^2 =$ 49
4	5	$5 - 7.167 =$ -2.187	-2.167	$-2.167^2 =$ 4.696	$(5-1)^2 =$ 16
5	10	$10 - 7.167 =$ 2.833	2.833	$2.833^2 =$ 8.026	$(10-1)^2 =$ 81
6	9	$9 - 7.167 =$ 1.833	1.833	$1.833^2 =$ 3.360	$(9-1)^2 =$ 64
7	12	$12 - 7.167 =$ 4.833	4.833	$4.833^2 =$ 23.369	$(12-1)^2 =$ 121
8	13	$13 - 7.167 =$ 5.833	5.833	$5.833^2 =$ 34.024	$(13-1)^2 =$ 144
9	7	$7 - 7.167 =$ $-.167$	$-.167$	$.167^2 =$ $.028$	$(7-1)^2 =$ 36
10	4	$4 - 7.167 =$ -3.167	-3.167	$3.167^2 =$ 10.030	$(4-1)^2 =$ 9
11	3	$3 - 7.167 =$ -4.167	-4.167	$4.167^2 =$ 17.364	$(3-1)^2 =$ 4
12	8	$8 - 7.167 =$ $.833$	$.833$	$.833^2 =$ $.694$	$(8-1)^2 =$ 49
13	9	$9 - 7.167 =$ 1.833	1.833	$1.833^2 =$ 3.360	$(9-1)^2 =$ 64
14	5	$5 - 7.167 =$ -2.167	-2.167	$-2.167^2 =$ 4.696	$(5-1)^2 =$ 16
15	1	$1 - 7.167 =$ -6.167	-6.167	$-6.167^2 =$ 38.032	$(1-1)^2 =$ 0
16	6	$6 - 7.167 =$ -1.167	-1.167	$-1.167^2 =$ 1.362	$(6-1)^2 =$ 25
17	9	$9 - 7.167 =$ 1.833	1.833	$1.833^2 =$ 3.360	$(9-1)^2 =$ 64
18	2	$2 - 7.167 =$ -5.167	-5.167	$-5.167^2 =$ 26.698	$(2-1)^2 =$ 1
19	14	$14 - 7.167 =$ 6.833	6.833	$6.833^2 =$ 46.690	$(14-1)^2 =$ 169
20	9	$9 - 7.167 =$ 1.833	1.833	$1.833^2 =$ 3.360	$(9-1)^2 =$ 64
21	4	$4 - 7.167 =$ -3.167	-3.167	$-3.167^2 =$ 10.030	$(4-1)^2 =$ 9
22	11	$11 - 7.167 =$ 3.833	3.833	$3.833^2 =$ 14.692	$(11-1)^2 =$ 100
23	8	$8 - 7.167 =$ $.833$	$.833$	$.833^2 =$ $.694$	$(8-1)^2 =$ 49
24	7	$7 - 7.167 =$ $-.167$	$-.167$	$-.167^2 =$ $.028$	$(7-1)^2 =$ 36
25	4	$4 - 7.167 =$ -3.167	-3.167	$-3.167^2 =$ 10.030	$(4-1)^2 =$ 9
26	7	$7 - 7.167 =$ $-.167$	$-.167$	$-.167^2 =$ $.028$	$(7-1)^2 =$ 36
27	9	$9 - 7.167 =$ 1.833	1.833	$1.833^2 =$ 3.360	$(9-1)^2 =$ 64
28	6	$6 - 7.167 =$ -1.167	-1.167	$-1.167^2 =$ 1.362	$(6-1)^2 =$ 25
29	5	$5 - 7.167 =$ -2.167	-2.167	$-2.167^2 =$ 4.696	$(5-1)^2 =$ 16
30	4	$4 - 7.167 =$ -3.167	-3.167	$-3.167^2 =$ 10.030	$(4-1)^2 =$ 9
	$\Sigma Y_i = 215$	$\Sigma(Y_i - \bar{Y}) =$ 0		$\Sigma(Y_i - \bar{Y})^2 = 304.181$	$(Y_i - 1)^2 = 1445$

vided *equally*. This suggests that the mean may be used to play an egalitarian role. The median, on the other hand, is the most *representative* score in the sense of being closest to every other score. This suggests that the median may be used to evaluate norm or standard violations by checking whether the observed median point falls above or below a norm or standard. The mode is the *most likely* score in the sense of being the most frequent. This suggests the mode may be used to represent the most common situation.

Given these interpretations, the choice of measures depends on the purpose of the research. Suppose that a welfare agency is to be evaluated. Previous experience suggests that the number of clients (an interval variable) each caseworker has on his or her caseload is an important criterion of effectiveness. Once the caseload size for each caseworker is obtained, each measure of central tendency should be calculated. Suppose that the caseload size ranged from 5 to 59 clients per caseworker and that the measures of central tendency were:

$$\overline{Y} = 29 \text{ clients per caseworker}$$
$$Md = 30 \text{ clients}$$
$$Mo = 40 \text{ clients.}$$

What do these measures indicate about the effectiveness of this agency?

The first question asked might be: what is the most common caseload for caseworkers in this agency? The *mode* shows that the most likely caseload is 40 clients per caseworker.

Previous experience suggests that caseworkers are most effective when their agency has them serve no more than 25 clients, which may be used as a norm or standard of effectiveness. The median of 30 clients per caseworker in this agency shows that over 50 percent of the caseworkers exceed this standard of effectiveness.

To allow caseworkers to be more effective, a consultant might then tell the agency it should redistribute clients among caseworkers to equalize the caseload. Will this equalization solve the problem in terms of effectiveness standard? No. The *mean* of 29 clients per caseworker indicates that equalization may improve the situation, but it will not bring down the caseload size to the desired 25 clients per caseworker. To solve the problem, then, other caseworkers would need to be hired, clients dropped, or the effectiveness standard changed.

The choice between measures of central tendency, then, depends on the criteria in the decision-making chart (Figure 7.2) and on the purpose of the research.

7.7 SPSS

Means, medians, and modes are calculated in SPSS through the FREQUENCIES program and the use of the STATISTICS card, as shown in

Section 5.9. Tables 5.11 through 5.14 report these measures for the variables in the detention home study. Note that, whether appropriate or not, all measures are reported for all variables. The computer cannot differentiate between nominal, ordinal, and interval variables (as you can), and therefore it calculates all measures. Report only the appropriate measures.

Chapter 8

Measures of Variability

Once a univariate distribution, a graph of its shape, and a measure of central tendency are present, the only other piece of information required to describe the distribution is a measure of variability.

8.1 DEFINITION

Measures of variability are single numbers that represent the extent of spread, scatter, or dispersion among scores in a distribution. Figure 8.1 shows two distributions differing in variability. The more narrow distribution (distribution B) has less variability than the flatter distribution (distribution A).

Measures of variability are often used to indicate how "typical" your measure of central tendency is. When there is small variability among scores, the distribution will be more narrow, and the measure of central tendency will be closer in value to all scores in the distribution. When there is large variability among scores, the distribution will be flatter, and the measure of central tendency will be less close in value to all scores in the distribution. Remember that a measure of central tendency summarizes a frequency distribution by a single number. If that summary measure were then used to predict or guess each score in the frequency distribution, the measure of variability would indicate the amount of error you would make in using a measure of central tendency as a typical score for all scores in a distribution. (In Chapter 11, this will be called *error of prediction*).

There are several measures of variability, each of which describes the extent of variability differently. The measures included in this book are the variation ratio, range, variance, and standard deviation. Choice among them depends on the scale of measurement for the variable and on the type of central tendency measure used to describe the typical score. Because the calculation of measures of central tendency must often be done before measures of variability may be computed, concern about the scale of measurement is implicit or understood. Figure 8.2, therefore, summarizes the choices to be made only on the basis of which measure of central tendency was calculated beforehand. If the mode was calculated,

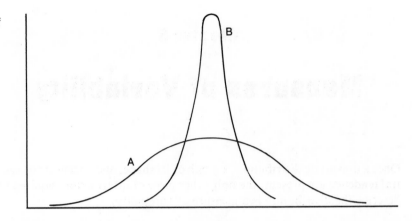

Figure 8.1.   Distributions differing in variability.

then the variation ratio is the appropriate measure of variability. If the median was calculated, then the range is the appropriate measure of variability. If the mean was calculated, then either the variance or standard deviation would provide an appropriate measure of variability.

Throughout this chapter, the interval variable "days spent" in the detention home study will be used to calculate the measures of variability.

8.2   VARIATION RATIO

The *variation ratio* is the proportion of subjects not in the modal category. The variation ratio is appropriate to use as a measure of variability when the mode is the measure of central tendency. The symbol for the variation ratio is v.

The variation ratio is calculated by the following formula:

$$v = 1 - \frac{f_{modal}}{n}$$

where

f_{modal} = the frequency of the modal category, which is the category where the mode is located

n = the total number of subjects.

This ratio, then, is 1 minus the proportion of subjects in the modal category. The closer the variation ratio is to 1.0, the greater the variability of scores and the less typical the mode is of all scores. The closer the variation ratio is to zero, the smaller the variability of scores and the more typical the mode is of all scores.

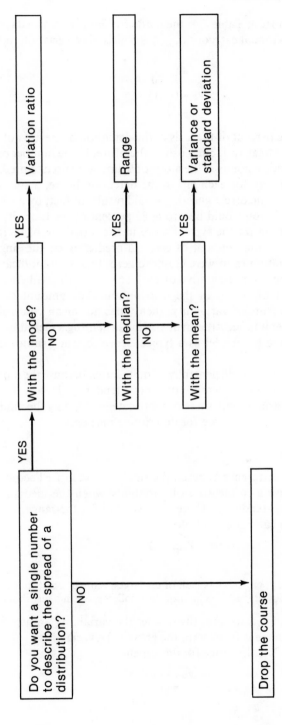

Figure 8.2.   Decision-making chart for measures of variability.

As is known from Table 7.1, the mode for days spent is 9 days. The frequency of this modal category (f_{modal}) is 5 girls. The variation ratio is:

$$v = 1 - \frac{5}{30}$$
$$= 1 - .17$$
$$= .83.$$

Eighty-three percent of the girls in the detention home are not in the modal category. Great variability exists, therefore, in the number of days spent in the detention home. If the mode of 9 days were used to predict the typical length of stay for each girl in the detention home, a substantial amount of error or incorrect guessing would result. In fact, by guessing 9 days for each girl, you would be wrong 83 percent of the time. By using the mode to summarize the typical score in the distribution, therefore, you make a substantial amount of error in predicting or guessing each score. If the variation ratio were closer to zero, this would indicate that the mode was more representative of each score and thus a better summary description of scores in the distribution. The amount of error reflected in a measure of variability, then, does not mean your subjects are in error; rather it reflects the error you make in using a measure of central tendency (here the mode) as a typical score for each subject in the distribution.

The variation ratio will generally be quite large, because the mode is a rather poor and unstable measure of central tendency. However, if you have a nominal variable with three or more categories, only the mode and variation ratio are appropriate for descriptive purposes.

8.3 RANGE

The *range* is the difference between the smallest and largest scores. It is appropriate to use as a measure of variability when the median is the measure of central tendency. There is no symbol for the range.

The range is calculated as follows:

$$\text{range} = Y_{largest} - Y_{smallest}$$

where

$Y_{largest}$ = the upper true class limit of the largest score in a distribution
$Y_{smallest}$ = the lower true class limit of the smallest score in a distribution.

The closer the range is to zero, the smaller the variability among scores. The further the range is from zero, the greater the variability.

From Table 7.1, the range for days spent is:

$$\text{range} = 14.5 - .5$$
$$= 14.$$

There is a range of 14 days that girls stay in the detention home.

Because the range only uses the end points of the scale, it tends to be very unstable from sample to sample, and it can tell you nothing about the scores in the middle of the distribution. The range is only used, therefore, when a quick and dirty measure of variability is needed.

Other measures exist to use with the median, such as semi-interquartile range (not discussed here because it is seldom used and is not in SPSS). The basic problem with all such alternative measures is that they do not give a clear sense of variability among scores. Moreover, few of them allow the use of more complex statistics. The pressure, then, is always to try to use interval-level statistics.

8.4 VARIANCE

The *variance* is the average of the squared deviations of scores about the mean. The symbol for the variance of a population is σ^2 (Greek sigma squared). The symbol of the variance of a sample is s^2. Another sample variance is the *biased* estimate of a population variance, symbolized by a capital S^2. The sample variance shown above is the *unbiased* sample variance (s^2). The distinction between biased and unbiased is discussed later. The important point here is that the biased estimate is calculated in the same way as the population variance, with N in the denominator, whereas the unbiased estimate has ($n-1$) in the denominator. The choice between the two estimates depends on whether the variance is used to describe only the data in the frequency distribution (use S^2), and thus the data are treated as population data (σ^2), or whether the variance is used to estimate or infer an unknown population variance (use s^2) on the basis of sample information. Henceforth, the only distinction made will be between σ^2 and s^2.

The definition of variance describes how it is calculated using a data list. However, the calculation differs somewhat for a population and for a sample. If the variance is to be used in a description of a *population* of subjects, then it is calculated as:

$$\sigma^2 = \frac{\Sigma(Y_i - \mu)^2}{N}.$$

If the variance is to be used in a description of a *sample* of subjects, with the purpose of estimating the unknown population variance, then it is calculated as follows:

$$s^2 = \frac{\Sigma(Y_i - \overline{Y})^2}{n-1}.$$

where

$Y_i =$ scores for all subjects, $i = 1$ through N or n
$\Sigma(Y_i - \mu)^2 =$ sum of squared deviation scores about the population mean
$\Sigma(Y_i - \overline{Y})^2 =$ sum of squared deviation scores about the sample mean
$N =$ total number of subjects in the population
$n =$ total number of subjects in the sample.

To understand the variance, the calculation of both the numerator and the denominator needs further explanation.

The *numerator* of the variance formula is based on the concept of variability and on the two properties of the mean (review Section 7.5). The concept of variability reflects the amount of spread or scatter about a measure of central tendency (in this case, the mean). If the scores of subjects are clustered closely around the mean of a distribution, then the extent of their spread or variability is small. If the scores of subjects are considerably scattered or distant from the mean of a distribution, then the extent of their variability is large. The distance of scores from the mean reflects, therefore, the extent of variability of scores about the mean. However, property 1 of the mean shows that, if such deviation scores about the mean were to be summed to get an overall measure of variability for the distribution, the result would always be zero regardless of the actual extent of scatter about the mean: $\Sigma(Y_i - \mu) = 0$. Property 2 of the mean, the least-squares property, may be used to get around this problem. This property states that the sum of squared deviations about the mean will be a minimum deviation value. Thus, each deviation score about the mean should be squared before adding all the deviation scores together. The amount of scatter about the mean, when calculated in this fashion, will always be a minimum value and yet will reflect the amount of variability in a distribution. The numerator of the variance formula, then, is the sum of squared deviations of scores about the mean of a distribution. As such, the numerator is a least-squares approach to defining variability.

Because "the sum of squared deviations about the mean" is cumbersome to write, it is often referred to as the *total sum of squares*. The numerator of the variance formula will often be called the total sum of squares henceforth.

The *denominator* of the variance formula is based both on the concept of an average or mean and on the concept of bias. By dividing the total sum of squares by the total number of subjects, you get the average of the squared deviations of scores about the mean, which by definition is the variance. In other words, the variance is the average squared deviation of all scores from the mean.

But why divide by N for a population and by $(n - 1)$ for a sample? A full explanation is beyond the scope of this book. Briefly, however, if sample variances are used to estimate unknown population variances, division by n rather than $(n-1)$ would give a *biased* result. "Biased" is a statistical term which means that, for randomly drawn samples of the same size, the average of these sample variances would *not* equal the population variance. For the variance, the values would be too small on the average by a factor of $(n-1)/n$. If $n=5$, the variance estimates would on the average be $(5-1)/5=.80$ or 80 percent of what they should be, and hence 20 percent too small. To compensate for this bias, $(n-1)$ rather than n is used as the divisor. This makes the sample variance, s^2, *unbiased* (in the long run, the average of many such randomly drawn sample variances would equal the population variance). Although this is an important theoretical point, it should be noted that if n is 100 or more, the effect of dividing by $(n-1)$ rather than by n is negligible. Practically, then, this last point is more important for small samples than for large.

To calculate the variance for days spent in the detention home, apply the formulas to the data list in Table 7.2. The numerator for both the population and the sample variance is 304.181, as calculated in column 4 of that table. If the 30 girls are treated as a population, the variance is:

$$\sigma^2 = \frac{304.181}{30}$$
$$= 10.139 \text{ square days.}$$

If the 30 girls are treated as a sample from a larger population, the variance is:

$$s^2 = \frac{304.181}{29}$$
$$= 10.489 \text{ square days.}$$

Note that the population and sample variances are not the same. This will always be the case. When the population variance is calculated, the 30 girls are considered to be the total population. Consequently, no sampling error is introduced in the calculation of the variance. When the sample variance is calculated, the 30 girls are considered to be a sample from some larger population of girls. If many randomly drawn samples of the same size (but not necessarily of the same girls) were observed, the average of all the sample variances would equal the variance for that larger population.

The variance is not intuitively easy to interpret because it is stated in "square" units of the measurement scale used. How many people would understand the variability of days spent, if described as 10.139 or 10.489 "square" days? Because the usual answer would be no one, the standard deviation is usually reported rather than the variance.

8.5 STANDARD DEVIATION

The *standard deviation* is the square root of the variance. As such, the standard deviation provides a measure of variability in the same units as those in which the variable is measured. It is appropriate to use as a measure of variability when the mean is used as the measure of central tendency. The symbol for the standard deviation of a population is σ (Greek sigma). The symbol for the standard deviation of a sample is s.

If the standard deviation is to be used in a description of a *population* of subjects, then it is calculated as:

$$\sigma = \sqrt{\frac{\Sigma(Y_i - \mu)^2}{N}}$$

If the standard deviation is to be used in a description of a *sample* treated as a population, it is calculated as σ above, and its symbol is S. However, if the standard deviation is used with the purpose of estimating an deviation, then it is calculated as:

$$s = \sqrt{\frac{\Sigma(Y_i - \overline{Y})^2}{n-1}}$$

Thus, in the detention home study, if the 30 girls are considered to be the population, the standard deviation is:

$$\sigma = \sqrt{\frac{304.181}{30}}$$
$$= \sqrt{10.139}$$
$$= 3.184 \text{ days.}$$

If the 30 girls are considered to be a sample from a larger population, the standard deviation for the estimated population is:

$$s = \sqrt{\frac{304.181}{29}}$$
$$= \sqrt{10.489}$$
$$= 3.239 \text{ days.}$$

Note that the population and sample standard deviations are not the same. This will always be the situation because the sample standard deviation only estimates (and thus need not equal) the standard deviation from some larger population. If many randomly drawn samples of the same size were observed, the average of all the sample standard deviations would closely approximate the standard deviation for that larger population.

Both the population and sample standard deviations, however, indicate the average deviation of a score from the mean. Thus, for example, if you guessed (or predicted) the mean as the typical number of days spent by each girl in the detention home, you would on the average make an error of 3.2 days on each guess.

If a frequency distribution is used to calculate the standard deviation rather than a data list, then apply this formula:

$$s = \sqrt{\frac{\Sigma f_j(Y_j - \overline{Y})^2}{n-1}}$$

where

$j =$ score intervals in the distribution, with $j =$ first interval through the last interval
$f_j =$ frequency of each score interval
$\overline{Y}_j =$ midpoint of each score interval
$\overline{Y} =$ sample mean
$n =$ total number of subjects.

When this formula is used to calculate the standard deviation for days spent in the frequency distribution with grouped intervals of 1 week in Table 5.6, the result is:

$$s = \sqrt{\frac{16(4 - 7.167)^2 + 14(11 - 7.167)^2}{29}}$$
$$= \quad 3.553 \text{ days.}$$

This result differs from the standard deviation calculated from a data list because, with grouped intervals, every subject's score must be estimated by the midpoint of the grouped interval. The larger the grouped interval, the greater the error. Whenever possible, then, standard deviations should be calculated from data lists, or from frequency distributions with ungrouped intervals.

8.6 INTERPRETATION OF THE STANDARD DEVIATION

In discussing the standard deviation initially, it was observed that the spread of scores was small if they were clustered closely about the mean and that the spread was large if the scores were dispersed over considerable distances away from the mean. It may now be said, correspondingly, that *if the standard deviation of a distribution is small, the scores are concentrated near the mean, and if the standard deviation is large, the scores are scattered widely about the mean.* This statement has been formalized in three ways: 1) Tchebysheff's theorem, 2) the empirical rule, and 3) the standard, normal distribution of z scores.

Tchebysheff's theorem states that *at least* $(1 - 1/k^2)100$ percent of scores in a distribution of any shape will lie within k standard deviations of their mean. For example, if $k = 2$ standard deviations, at least 75 percent or $(1-1/2^2)100$ of the scores will lie within 2 standard deviations of their mean. For the days spent distribution in Table 7.1, this means that at least 75 percent of the girls must fall between .689 days ($\overline{Y}-2s = 7.167-2$ (3.239)) and 13.645 days ($\overline{Y}+2s = 7.167-2(3.239)$). Because almost all girls fall within 2 standard deviations of the mean, Tchebysheff's interpretation of *at least 75 percent* is correct. In most cases, the percent of scores contained within such an interval will be substantially in excess of Tchebysheff's theorem. The empirical rule often provides a better approximation and is thus used much more frequently.

The *empirical rule* applies exactly to scores that are normally distributed (distributions which are symmetric, unimodal, and mesokurtic), but also gives fairly good approximations to relatively normal but not perfectly normal distributions. The empirical rule states that *approximately:*

> 68.3 percent of the scores will fall within 1 standard deviation of the mean ($\overline{Y} \pm 1s$)
> 95.4 percent of the scores will fall within 2 standard deviations of the mean ($\overline{Y} \pm 2s$)
> 99.7 percent of the scores will fall within 3 standard deviations of the mean ($\overline{Y} \pm 3s$).

For the "days spent" distribution in Table 7.1, this means that approximately 95 percent of the girls will fall within 2 standard deviations from the mean or between 0.689 days and 13.645 days. This approximation is much closer to reality than Tchebysheff's estimate.

Where do these approximations for the empirical rule come from? The following sections on standard scores and standard normal scores provide an answer.

8.7   STANDARD SCORES

A *standard score* indicates the number of standard deviations a score is above or below the mean. The symbol for a standard score, for both a population and a sample, is z. A standard score is often called a *z score*.

If a standard score is used to describe a score in a *population*, then it is calculated as:

$$z = \frac{Y_i - \mu}{\sigma}.$$

If a standard score is used to describe a score in a *sample*, then it is calculated as:

$$z = \frac{Y_i - \overline{Y}}{s}.$$

where

Y_i = scores for subjects
μ = population mean
$\overline{Y}$ = sample mean
$(Y_i - \mu)$ = deviation score of a subject's score from the population mean
$(Y_i - \overline{Y})$ = deviation score of a subject's score from the sample mean
σ = population standard deviation
s = sample standard deviation.

For example, in a sample, the standard score for 12 days spent would be:

$$z = \frac{12 - 7.167}{3.239}$$
$$= 1.492 \text{ standard deviations.}$$

The standard score for 3 days spent would be:

$$z = \frac{3 - 7.167}{3.239}$$
$$= -1.286 \text{ standard deviations.}$$

There are several considerations in interpreting standard scores. First, transforming all scores into standard scores *does not* change the shape of the original frequency distribution, just as changing from inches to yards does not change the height of a building. If a distribution is skewed and bimodal before its scores are transformed into standard scores, it will be just as skewed and bimodal afterwards.

Second, a positive standard score indicates that the score is above the mean. The standard score for 12 days is positive, therefore, because 12 days is above the mean of 7.167 days. A negative standard score indicates that the score is below the mean. The standard score for 3 days is negative, therefore, because 3 days is less than the mean of 7.167. When a score is the same value as the mean, the standard score will be zero.

Third, a standard score represents the distance of a score from a mean $(Y_i - \overline{Y})$ in standard deviation (s) units of measurement. The standard score of 1.492 for 12 days indicates that 12 days, or the deviation score of 4.833 days (12 − 7.167), is 1.492 standard deviations (4.833/3.239) *above* the mean. The standard score of −1.286 for 3 days indicates that 3 days or the deviation score of −4.167 days (3 − 7.167) is 1.286 standard deviations (−4.167/3.239) *below* the mean. Division of the distance of a score from a mean, the deviation score, by the standard deviation provides a procedure for transforming scores from their original unit of measurement (days spent) to a common statistical unit of measurement (standard deviations from the mean).

Standard scores, then, provide a common yardstick (number of standard deviations from the mean) to indicate the relative position of scores in a distribution. If, for example, a welfare mother had seven children, and the group mean was six children ($\overline{Y} = 6$) and the standard deviation was one child ($s = 1$), then she is one standard deviation above the group mean for number of children. Although the absolute number of seven children provided no information on relative group position for this mother, the standard score does indicate the relative position of this mother as being a standard deviation above the mean.

Standard scores may also be used to add scores from distributions with different means, standard deviations, and units of measurement. To get a measure of social class, for example, it is often necessary to add a subject's scores for education, income, and occupation. Each of these variables is measured differently. Thus, they may only be added legitimately by first transforming each score into a standard score or common unit of measurement. When this is done, the distributions for education, income, and occupation will all have the same mean (zero), the same standard deviation (1.00), and the same unit of measurement (number of standard deviations from the group mean). The use of standard scores is thus like adding apples to apples, whereas the use of original or raw scores here would be like adding apples to oranges.

Fourth, as mentioned above, the mean of standard scores for all distributions is zero. Recall that property 1 of the mean (review Section 7.5) states that the sum of deviation scores about the mean is always zero. Since the numerator of the standard score formula is a deviation score ($Y_i - \overline{Y}$), the sum of these deviation scores will always be zero. Thus, the sum of standard scores will always be zero, as will the mean of standard scores (a zero divided by any number, including either a standard deviation or a sample size, will always be zero). For example, say there are three scores in a sample distribution (1, 3, and 5). The mean of this distribution is 3, and the sample standard deviation is 2. The deviation scores for each score are -2, 0, and 2; the sum of these is zero. The standard scores for each score are -1 ($-2/2$), 0 ($0/2$), and 1 ($2/2$); the sum of these is zero. The mean of deviation scores ($0/3$) and the mean of standard scores ($0/3$) is zero. This may also be shown in symbols as follows. The mean of standard z scores is:

$$\overline{Y}_z = \frac{\Sigma z_i}{n} = \frac{\Sigma \left(\dfrac{Y_i - \overline{Y}}{s} \right)}{n} = \frac{\Sigma (Y_i - \overline{Y})}{ns} = \frac{1}{ns} \Sigma (Y_i - \overline{Y});$$

and because, under property 1 of the mean, $\Sigma(Y_i - \overline{Y}) = 0$:

$$\overline{Y}_z = \frac{1}{ns}(0) = 0.$$

That is, the mean of standard scores is zero.

Fifth, the standard deviation and variance of standard scores for all distributions is 1.00. In the above example of three scores (1, 3, and 5) and their corresponding standard scores (-1, 0, and 1), the variance and standard deviation of the standard scores is 1.00. Note that if the variance is always 1.00, so must the standard deviation be equal to 1.00, because the square root of one is one. This may be shown in symbols as follows. The variance of standard scores is:

$$S_z^2 = \frac{\Sigma(z_i - 0)^2}{n-1} = \frac{\Sigma z_i^2}{n-1} = \frac{\Sigma\left(\frac{Y_i - \overline{Y}}{s}\right)^2}{n-1} = \frac{\Sigma(Y_i - \overline{Y})^2}{n-1} \cdot \frac{1}{s^2}.$$

and since the variance of raw scores is:

$$s^2 = \frac{\Sigma(Y_i - \overline{Y})^2}{n-1},$$

the last term may be reduced to 1.00:

$$s_z^2 = \frac{\Sigma(Y_i - \overline{Y})^2}{n-1} \cdot \frac{1}{s^2} = \frac{s^2}{1} \cdot \frac{1}{s^2} = \frac{s^2}{s^2} = 1.00.$$

The standard deviation is also equal to one because $s = \sqrt{s^2} = \sqrt{1.00} = 1.00$.

8.8   STANDARD NORMAL SCORES

A *standard normal score* indicates the number of standard deviations a score is above or below the mean of a normal distribution of scores. A standard normal score shares the same symbol as a standard score (z), and it is calculated in the same manner. The one great difference between them is that standard normal scores are always normally distributed. That is, the distribution of their scores is bell-shaped as well as symmetric, unimodal, and mesokurtic. Figure 8.3 shows a normal distribution. The normal distribution will be discussed more in Part III, but it is of interest now because it is the basis of the empirical rule.

Recall that the area contained within a graph or curve represents the total sum of proportions (1.00) or percentages (100%) of a distribution. The area under the standard normal curve in Figure 8.3 may also be interpreted in this way. The height of the curve represents the proportion, percentage, or likelihood of a score. The normally distributed scores are placed along the X axis; here the scores are z scores.[1] A zero represents the mean: the most typical score or mean in a distribution of z scores and thus the score under the highest part of the curve. The positive numbers to the right of the zero are z scores: $+1$ indicates a score is 1 standard deviation unit above the mean, $+2$ indicates a score is 2 standard deviation units above the mean, $+3$ indicates a score is 3 standard deviation units above the mean, and so on. The negative numbers to the left of the zero are also z scores: -1 indicates a score is 1 standard deviation unit below the mean, -2 indicates a score is 2 standard deviation units below the mean, -3 indicates a score is 3 standard deviation units below the mean, and so on. Look at Figure 8.3 again: z scores between -1 and $+1$ include exactly 68.3 percent of all the scores for subjects, scores between -2 and $+2$ include exactly 95.4 percent of all scores, and scores between -3 and $+3$ include exactly 99.7 percent of all scores.

Table B in the Appendix provides these exact proportions or percentages for every possible z score that is normally distributed. The first column is a z score to the first decimal place. The top row adds further precision by giving a z score to two decimal places. For example, a z score of 1.00 is found by looking down the first column until 1.0 is found. Since the second decimal place is zero, the first top row labeled .00 is correct. The proportion of area between the mean and a z score of 1.00 within this table is .3413. Thus 34.13 percent of subjects' scores are between the mean and 1 positive standard deviation unit. Because a normal distribution, and therefore a standard normal distribution, is symmetrical, the same proportion of subjects' score lies between the mean and 1 standard deviation below the mean. Thus, 68.3 percent of subjects' scores lie between -1 and $+1$ standard deviations from the mean ($.683 = .3413 + .3413$). Repeat this process for ±2 and ±3 standard deviations from the mean and it will be seen that the empirical rule emerges. Moreover, the closer a distribution of scores comes to being a normal distribution, the more exact the empirical rule becomes. If the distribution of scores is normal, then the word "approximately" in the empirical rule is changed to "exactly." In a normal distribution, therefore, *exactly*:

> 68.3 percent of the scores will fall within 1 standard deviation of the mean ($\overline{Y} \pm 1s$)
>
> 95.4 percent of the scores will fall within 2 standard deviations of the mean ($\overline{Y} \pm 2s$)
>
> 99.7 percent of the scores will fall within 3 standard deviations of the mean ($\overline{Y} \pm 3s$).

---

[1]The scores used are z scores because all raw scores may be transformed to z scores with a mean of zero and a standard deviation of one. Thus, only one table of areas under the normal curve for z scores (see Table B in the Appendix) is necessary for any normal distribution. Otherwise, hundreds of tables for every different combination of means and standard deviations would be required for distributions of raw scores.

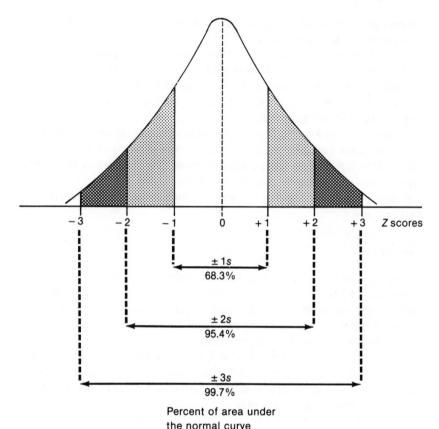

Percent of area under
the normal curve

Figure 8.3.   A standard, normal curve.

The word "exactly" is the reason the normal distribution and curve have been of tremendous applied importance in statistics. Part III will discuss this point further.

8.9   USE

The standard deviation is used much more than other measures of variability. The standard deviation is desired because of its many applications.

First, the standard deviation is used to calculate many other statistics.

Second, the standard deviation may be used to compare the variability of scores between groups. If two welfare agencies have the same average caseload per caseworker, but one has a standard deviation of 3.0 clients and the other has a standard deviation of 15.0 clients, then the second agency has much more variation in its caseload size than the first agency. A consultant would probably have much more trouble getting the second agency to equalize its caseload (because it would have to change more) than the first agency.

Third, if scores are normally distributed or even approximately so, then the empirical rule may be applied to describe the percentage of subjects who fall within 1, 2, or 3 or more standard deviations of the mean. If scores do not follow a normal distribution, then Tchebysheff's theorem may be used for this purpose.

Last, use of the standard deviation in the calculation of standard normal scores allows an interpretation of each score in terms of its number of standard deviation units from the group mean. This procedure provides a perspective on the relative position of a single score in a distribution and the relative position of two or more scores for a single subject in different distributions.

8.10 SPSS

The range, variance, and standard deviation are calculated in SPSS through the FREQUENCIES program and the use of the STATISTICS card, as shown in Section 5.9. Tables 5.11 through 5.14 report these measures for the variables in the detention home study. Several observations should be noted about these tables.

First, all measures are reported for all variables, because the computer cannot differentiate among nominal, ordinal, and interval variables. You should report only the appropriate measures using the criteria established in this chapter.

Second, the variance and standard deviation are calculated with $(n-1)$ in their denominators and thus correspond to the sample statistic formulas presented here.

Third, the variation ratio is not calculated in SPSS. If this measure of variability is desired, it must be done by hand calculation.

Fourth, the range is calculated differently in SPSS. SPSS simply subtracts the smallest score (minimum) from the largest score (maximum) without regard to their true class limits. The result is intuitively clearer, however, if the range is calculated as discussed in Section 8.3.

Fifth, the only SPSS statistic not yet discussed is the standard error (abbreviated "Std Err"). This statistic will be defined and discussed in Part III.

Finally, it should be noted that Chapter 8 of the *SPSS Manual* also discusses the CONDESCRIPTIVE program for describing interval-level variables only. This program also computes z scores, if the third option in the program is chosen. The z scores are computed by the sample formula with s in the denominator.

At this point, the entire FREQUENCIES printout should be under your control and understanding. Congratulations.

Chapter 9

Bivariate Frequency Distributions

Previous chapters focused on the description of one variable at a time. Often, however, interest is focused on the description of two variables together. Interest in a bivariate analysis usually arises because there is a question about whether income, for example, is distributed differently for men and women. In other words, is income related to sex? There is an endless number of such questions: Is intelligence related to age? Is drug addiction related to social class? Is agency effectiveness related to the number of its caseworkers? Is mental illness related to crime? and so on. The focus of the remaining chapters in Part II is on how to describe such bivariate relationships.

There are many ways to describe a bivariate relationship. Figure 9.1 summarizes the major choices. If all information about both variables together is needed, then either a contingency table of their joint distribution (this chapter) or a bivariate graph (Chapter 10) should be made. If a single number to characterize their relationship is needed, then a measure of relationship should be the choice (Chapter 11).

No matter which choice is made, however, there are four descriptive characteristics of a relationship:

1. whether or not a relationship *exists*
2. the *strength* of that relationship
3. the *direction* of the relationship
4. the *type* of the relationship.

Each characteristic will be discussed separately in this chapter. In Chapter 11, measures of relationship are used that describe at least the first two and sometimes all four characteristics in a single number.

9.1 DEFINITION

A *bivariate frequency distribution* is the joint distribution of two variables that shows the frequency with which the score intervals of both variables co-vary with each other.

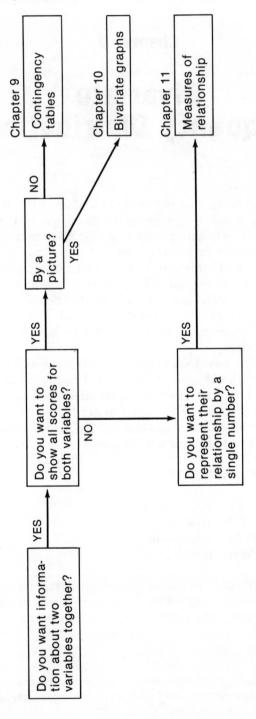

Figure 9.1.   Decision-making chart for methods to describe bivariate relationships.

A bivariate frequency table, often called a *contingency table*, is the method used to show the joint distribution of two variables. The score intervals of one variable form the columns of this table, and the score intervals of the other variable form the rows. The cells of a contingency table represent the intersection of score intervals of both variables, and the cell frequencies show the number of subjects who have both score intervals in common. The pattern of frequencies across cells indicates whether or not the score intervals of both variables co-vary. If they do co-vary, a relationship is said to exist between the two variables.

9.2 EXAMPLE

An example of a bivariate contingency table is given in Table 9.1. This table shows the joint distribution of two variables from the detention home study, seriousness of offense and number of days spent in the detention home. Seriousness of offense is located in the columns of Table 9.1. Each of the three columns represents one of its score intervals: low, moderate, and high seriousness of offense. Number of days spent in the detention home is located in the rows. Each of the two rows represents one of its score intervals: 1–7 days or 8–14 days spent in the home. Although days spent originally had 14 score intervals, with each interval representing the exact number of days spent in the detention home, these intervals have been collapsed into only two broader intervals so the table would not be too big. Another reason for collapsing score intervals may be to avoid having empty cells, that is, cells with no frequencies.

Each *cell* or box of a contingency table always indicates the joint occurrence or intersection of score intervals for the two variables. In Table 9.1, the upper lefthand cell represents the joint occurrence of girls who both committed offenses of low seriousness and spent 1–7 days in the detention home.

Each *cell frequency* represents the number of times these score intervals occur jointly for subjects. Each cell frequency is obtained by a joint count or *cross-tabulation* of scores from a data list. The cell frequencies in Table 9.1 are obtained from the information in the data list provided in

Table 9.1. Contingency table of seriousness of offense (independent variable) and days spent (dependent variable) for girls in detention home ($n = 30$)

| Days spent | Seriousness of offense | | | |
	Low	Moderate	High	Total
1–7	9 (75%)	5 (56%)	2 (22%)	16
8–14	3 (25%)	4 (44%)	7 (78%)	14
Total	12	9	9	30

Column percentages are given in parentheses.

Table 4.1. The frequency of the upper lefthand cell is obtained by counting the number of times a girl or subject on this data list both committed an offense of low seriousness and spent 1–7 days in the detention home. Nine girls fulfilled this joint criterion: subjects 1, 4, 9, 10, 15, 16, 26, 28, and 29. The cell frequency is therefore nine girls. This procedure is used to count the frequency of each cell.

Notice in Table 9.1 that the column and row score intervals both begin with the lowest value in the upper lefthand cell: offenses of low seriousness and 1–7 days spent. It is customary to start column and row score intervals of a contingency table in this way because it facilitates calculation and interpretation. This custom, however, only applies for ordinal and interval variables, because the concept of ranking from low to high is only appropriate for them. If a nominal variable is used in a contingency table, then the order of score of its intervals makes no difference.

Also notice that the row totals to the righthand side of the table and the column totals below the table for each score interval in Table 9.1 are exactly the same as the univariate frequency distributions for each variable alone (see the univariate frequency distributions for seriousness of offense and days spent, in Tables 5.2 and 5.5, respectively). These totals are called *marginal frequencies* or *marginals* because they are located in the margins of the contingency table.

Last, notice the title for Table 9.1 specifies what type of table it is (contingency), the variables in the table (seriousness of offense and days spent), the subjects (girls in a detention home), the number of subjects ($n = 30$), and what if anything is reported in parentheses (percentages). All contingency tables should be titled in this way.

9.3   USE

Contingency tables are used to detect a relationship between two variables. If both variables are either ordinal or interval, contingency tables may also be used to determine the direction and type of relationship.

A *relationship* is said to exist if two variables co-vary with each other. Co-vary means that the score intervals of both variables change or vary in some fashion together. That is, a change in one is often associated with a change in the other. When two variables co-vary, the score intervals of one variable are said to be associated with or to predict certain score intervals of another variable.

Moreover, in statistics, a relationship has two distinct levels of interpretation: within the sample itself or in the larger population. Within the sample itself, evidence of a relationship refers only to the sample information or data. In the larger population, a relationship found in the sample data may provide evidence that it also exists in the population; that is, an inference is made on the basis of evidence of the relationship in the sample. The difference of meaning is based on the distinction between descriptive and inferential statistics. Until Part III, where inferential statistics are discussed, the word "relationship" will only refer to the first meaning or to the existence of a relationship in a sample, with no inferences made about the existence of a relationship in the larger population.

There are two general ways of detecting a relationship: 1) measures of relationship, or 2) frequency, percentage, or proportion differences in a contingency table. Measures of relationship are the subject of Chapter 11. Percentage or proportion differences based on joint frequencies are discussed here as an intuitive "eyeball" method for detecting relationships.

Use of *percentage differences* for describing the existence of a relationship first requires the identification of independent and dependent variables. The independent variable of a contingency table is identified either by hypotheses derived from models and theories, or by the common-sense principle of ordering variables in terms of a time sequence. In both cases, the *independent variable is believed to predict the dependent variable*. Thus, an independent variable is often called the *predictor* and the dependent variable the *criterion* to be predicted. In Table 9.1, for example, seriousness of offense is identified as the independent variable because it precedes the other variable, number of days spent in the detention home, in terms of a time sequence. Seriousness of offense, then, precedes and is thus presumed to predict days spent. Note, as in Table 9.1, it is a convention to place the score intervals of the independent variable in the columns, and the score intervals of the dependent variable in the rows of a contingency table.

Once the independent variable is identified, *percentages are calculated in the direction of the independent variable.* If the score intervals of the independent variable are in the columns of the table, then column percentages are calculated. Each column, therefore, is treated just like a univariate percentage distribution. The percentages for each column or score interval of the dependent variable should thus sum to 100 percent. The "highly serious offense" column in Table 9.1, for example, contains nine girls, 22 percent of whom spent 1–7 days and 78 percent of whom spent 8–14 days in the detention home, for a total of 100 percent.

To see whether a relationship exists, differences between column percentages should be compared across the score intervals of the dependent variable (the rows in Table 9.1). *If there is any percentage difference between the column percentages for any row, then a relationship is said to exist in the sample.* If there is no relationship between variables, then all the percentage differences are zero or close to zero. *As the strength of a relationship increases, or the better the independent variable predicts the dependent variable, percentage differences will become larger and closer to 100 percent.* Yet judging the strength of a relationship in this way is very much an art, because percentage differences in a table may not be able to obtain a zero or 100 percent difference. The original distribution of each variable, shown in the margins of a contingency table, constrains the range of possible percentage differences. Thus, although percentage differences can indicate the existence of a relationship, they only suggest its relative strength. However, the measures of relationship to be discussed in Chapter 11 will provide a more exact and general way to evaluate strength. Until then, percentage differences allow the analysis of a bivariate distribution to show the existence and apparent strength of a relationship, albeit imprecisely.

These percentage difference criteria may be used to detect relationships and their apparent strength among all variables in the detention home study. In Table 9.1, for example, there are large percentage differences between low and high seriousness of offense for both 1–7 days ($75\% - 22\% = 53\%$) and for 8–14 days ($78\% - 25\% = 53\%$). These percentage differences indicate that a relationship exists between seriousness of offense and days spent and that it may be moderately strong. In Table 9.2, there are somewhat smaller percentage differences between 10–13 years of age and 14–17 years of age for both 1–7 days ($73\% - 33\% = 40\%$) and 8–14 days ($67\% - 27\% = 40\%$). Thus, although a relationship exists between age and days spent, it appears to be less strong. In Table 9.3, some very large percentage differences exist, as, for example, between 10–13 years of age and 14–17 years for low seriousness of offense ($73\% - 7\% = 66\%$). An apparently strong relationship exists, therefore, between age and seriousness of offense. In Tables 9.4 and 9.5, the small number of subjects in the "other" religion column suggest that percentage differences should only be based on comparisons of the columns for Protestant and Catholic (review Section 5.5). The resulting percentage differences

Table 9.2.   Contingency table of age (independent variable) and days spent (dependent variable) for girls in detention home ($n = 30$)

Days spent	Age		Total
	10–13 years	14–17 years	
1–7	11 (73%)	5 (33%)	16
8–14	4 (27%)	10 (67%)	14
Total	15	15	30

Column percentages are given in parentheses.

Table 9.3. Contingency table of age (independent variable) and seriousness of offense (dependent variable) for girls in detention home ($n = 30$)

Seriousness of offense	Age 10–13 years	Age 14–17 years	Total
Low	11 (73%)	1 (7%)	12
Moderate	3 (20%)	6 (40%)	9
High	1 (7%)	8 (53%)	9
Total	15	15	30

Column percentages are in parentheses.

suggest that if a relationship exists, it is very weak or close to zero. In Table 9.4, for example, a comparison between Protestant and Catholic girls shows that the percentage differences are quite small for both 1–7 days (60% − 46% = 14%) and 8–14 days (54% − 40% = 14%).

A relationship may also be said to have a *positive or negative direction*, if both variables are at least ordinal or interval. A *positive* relationship means that both variables increase (or decrease) in value together. Weight tends to increase as height increases, or weight tends to decrease as height decreases; thus there is a positive relationship between weight and height. A *negative* relationship means that, as one variable increases in value, the other variable decreases. Home heating bills increase, for example, as the temperature outside decreases. A negative relationship, therefore, exists between home heating bills and the outside temperature.

To determine the direction of an existing relationship, first locate the highest percentage in each score interval (each column) of the independent variable. Then determine the general pattern of these highest percentages. If the pattern tends to match low with low and high with high score intervals on both variables, then the relationship is *positive*. If the pattern tends to match low with high and high with low score intervals on both variables, then the relationship is *negative*. In Table 9.1, for example, the highest percentage for low seriousness of offense is 75 percent, for

Table 9.4. Contingency table of religion (independent variable) and days spent (dependent variable) for girls in detention home ($n = 30$)

Days spent	Religion Protestant	Catholic	Other
1–7	9 (60%)	5 (46%)	2 (50%)
8–14	6 (40%)	6 (54%)	2 (50%)
Total	15	11	4

Column percentages are in parentheses.

Table 9.5.   Contingency table of religion (independent variable) and seriousness of offense (dependent variable) for girls in detention home ($n = 30$)

Seriousness of offense	Religion			
	Protestant	Catholic	Other	Total
Low	6 (40%)	5 (46%)	1 (25%)	12
Moderate	5 (33%)	2 (18%)	2 (50%)	9
High	4 (27%)	4 (36%)	1 (25%)	9
Total	15	11	4	30

Column percentages are in parentheses.

moderate seriousness of offense it is 56 percent, and for high seriousness of offense it is 78 percent. The general pattern of these percentages is positive; that is, the pattern matches low with low and high with high score intervals on both variables. As seriousness of offense increases, length of stay in the detention home also increases. There is a positive relationship, therefore, between these two variables.

The relationships between other pairs of variables in the detention home study may be similarly analyzed. Table 9.2 shows a positive relationship between age and number of days spent in the detention home. As age increases, the number of days spent in the detention home increases. Table 9.3 shows a positive relationship between age and seriousness of offense. As age increases, the seriousness of offenses increases. However, because Tables 9.4 and 9.5 include a nominal variable, religion, it would make no sense to consider the direction of their relationships.

A final consideration is the *type of a relationship*. As with the above distinction between positive and negative relationships, this distinction only applies to relationships between ordinal or interval variables. Type of a relationship refers to the general trend of data in a table. A relationship is said to be *linear* (if both variables are interval) or *monotonic* (if one of the variables is ordinal), if an increase of one score interval in one variable tends to move *uniformly* up (or down) with changes in the score intervals of the other variable. A straight line, then, best represents the trend. Table 9.2 shows a linear relationship (two interval variables), and Tables 9.1 and 9.3 show monotonic relationships (one ordinal and one interval variable). If a curve best represents the trend, then the relationship is said to be *curvilinear*. If neither a straight line nor a curve fits the trend, the relationship is said to be *nonlinear*.

9.4   INTERPRETATION

Generally speaking, a relationship may be interpreted in three ways: 1) as cause and effect, where changes in the independent variable cause changes in the dependent variable; 2) as influence and outcome, where changes in the independent variable influence but do not necessarily cause changes in the dependent variable; and 3) as predictor and criterion, where changes in the independent variable are not known to cause, but are simply associated with and thus predict, changes in the dependent variable. Because the same measure of relationship or association is often used to support all three situations, choice of interpretation is based primarily on the type of research design incorporated and not on the type of statistical technique used. (Although a discussion of research design is beyond the scope of this book, generally speaking a research design is the researcher's specified and replicable method for collecting data, manipulating the intervention, or changing values of the independent variable, and controlling for all alternative explanations of the results.[1] This method determines the extent to which inferences may be made concerning the causality and generalizability of the results.) In other words, the presence of an empirical relationship or association does *not* indicate causation. It can only support causal arguments if the research design is appropriate. Suffice it to say that, unless indicated otherwise, all relationships discussed in this book should be interpreted in the predictor-criterion fashion. For example, although seriousness of offense is related to days spent in the detention home, one cannot say the seriousness of offense causes or influences length of stay, but only that seriousness of offense predicts length of stay. This rule of thumb applies to most research reports you might read as well. If a research design allows other than a predictor-criterion interpretation, the researcher will be sure to state it most clearly. Notice that emphasis must be placed on research design concerns and not just on statements from the researcher's theory. A researcher may offer a cause-and-effect argument on the basis of theory, but this does not make an empirical relationship a causal relationship *unless* the appropriate research design has been used.

9.5 SPSS

Contingency tables are made by SPSS in the CROSSTABS program. All the features of this program are discussed in Chapter 9 of the *SPSS Primer.*

---

[1]An excellent recent introduction to research design in clinical practice is John S. Wodarski, *The Role of Research in Clinical Practice: A Practical Approach for the Human Services* (Baltimore: University Park Press, 1981). A more general discussion and the classic introduction to research design is Donald T. Cambell and J. C. Stanley, *Experimental and Quasi-experimental Designs for Research* (Chicago: Rand McNally, 1967).

I used the following set of SPSS cards to get the contingency tables for the detention home study:

```
1                                16
RUN NAME                          ⎫
DATA LIST                         ⎪
INPUT MEDIUM                      ⎬  same as in Section 5.9
N OF CASES                        ⎪
VALUE LABELS                      ⎪
VAR LABELS                        ⎭
RECODE                           DAYS (1 THRU 7 = 1) (8 THRU 14 = 2)/
                                 AGE (10 THRU 13 = 10) (14 THRU 17 = 14)/
CROSSTABS                        TABLES  = DAYS BY OFFENSE/
                                          DAYS BY AGE/
                                          DAYS BY RELIGION/
                                          OFFENSE BY AGE/
                                          OFFENSE BY RELIGION/
STATISTICS                       ALL
READ INPUT DATA
  [place data cards here]
FINISH
```

The RUN NAME card through the VAR LABELS card have been discussed in Section 5.9, as have the READ INPUT DATA and FINISH cards.

The RECODE card allows the grouping of score intervals and is discussed in Chapter 6 of the *SPSS Primer*. Because both DAYS and AGE had so many original score intervals, they were collapsed into two score intervals each. DAYS was recoded to have two 1-week intervals. AGE was recoded to have an interval from 10 through 13 years and an interval from 14 through 17 years of age. As many variables may be recoded as needed by following the conventions of the RECODE card. Start at column 16 and punch the name of the variable you want recoded, as was done for DAYS. The first set of parentheses indicates the score intervals to be collapsed into the first week interval, and the second set of parentheses indicates the score intervals to be collapsed into the second week interval. The original score intervals are put inside the parentheses first, then are followed by an equals sign, and then any number is used to label the new grouped interval. A slash is put after the *last* score interval parenthesis to indicate that you have finished recoding that variable.

The CROSSTABS card tells the computer which variables should be put into contingency tables. "TABLES = " should be put before the list of tables. Variables to be put in a contingency table come afterward, with the name of the row variable first, then a space and the word BY, and then another space and the name of the column variable followed by a slash. For example, DAYS BY OFFENSE will produce a contingency table with DAYS as the row variable and OFFENSE as the column variable as shown in Table 9.6.

Table 9.6.   SPSS contingency table and statistics for age and days spent

	AGE		
Count Row % Col % Total %	10.	14.	Row Total
DAYS			
1.	11 68.8 73.7 36.7	5 31.3 33.3 16.7	16 53.3
2.	4 28.6 26.7 13.3	10 71.4 66.7 33.3	14 46.7
Column Total	15 50.0	15 50.0	30 100.0

Corrected chi-square = 3.34821 with 1 degree of freedom. Significance = 0.0673
     Raw chi-square = 4.82143 with 1 degree of freedom. Significance = 0.0281
Phi = 0.40089
Contingency coefficient = 0.37210
Lambda (asymmetric) = 0.35714 with DAYS dependent.
     = 0.40000 with AGE dependent.
Lambda (symmetric) = 0.37931
Uncertainty coefficient (asymmetric) = 0.11971 with DAYS dependent.
     = 0.11932 with AGE dependent.
Uncertainty coefficient (symmetric) = 0.11952
Kendall's Tau b = 0.40089   Significance = 0.0154
Kendall's Tau c = 0.40000   Significance = 0.0154
Gamma = 0.69231
Somers's D (asymmetric) = 0.40000 with DAYS dependent.
     = 0.40179 with AGE dependent.
Somers's D (symmetric) = 0.40089
Eta = 0.40089 with DAYS dependent.   = 0.40089 with AGE dependent.
Pearson's R = 0.40089   Significance = 0.0141

No OPTIONS card was included here because none were wanted. The STATISTICS card asks the computer to calculate "ALL" the statistics specified on page 79 of the *SPSS Primer*. Most of the statistics printed by the program here will be discussed in this book in Chapter 11. The exception is chi-square, a test to be discussed in Part III. As always, the computer will calculate statistics for any set of variables. Report only appropriate statistics.

Table 9.6 shows an example of a contingency table printed by SPSS, where DAYS is the row variable and AGE the column variable. Note that: 1) the names of the column and row variables are printed at the top and lefthand side of the table, 2) each row and column is labeled by the number representing its score interval, with the lowest numbered interval

beginning in the upper lefthand corner of the table, 3) the univariate frequency and percentage distributions for each variable are printed in the
margins of the table, 4) the joint frequency of the variables (Count) of
each cell is the top number in the cell, 5) the row percentage (Row %) is
the second number in the cell, 6) the column percentage (Col %) is the
third number in the cell, and 7) the total percentage (Total %) is the last
number in the cell and is based on the total sample size, n. Since AGE is
our independent variable, only the column percentages are directly useful
(see Table 9.2).

The requested statistics are printed underneath the contingency
table.

Chapter 10

Graphs of Bivariate Frequency Distributions

Bivariate graphs provide a picture of a relationship. There are many different graphs for bivariate distributions or relationships, but only the most common techniques will be illustrated here.

Figure 10.1 summarizes the choices between graphs. If both of your variables are either ordinal or nominal, then a multiple bar graph is the appropriate choice. If one variable is interval and the other is either ordinal or nominal, then multiple histograms and multiple polygons are appropriate. If both variables are interval, then a scattergram is the preferred choice.

A bivariate frequency or percent distribution is necessary beforehand to draw multiple bar graphs, histograms, and polygons. A data list of two variables, however, is necessary to draw a scattergram.

10.1 MULTIPLE BAR GRAPHS

A *multiple bar graph* consists of two or more sets of bar graphs, as defined in Section 6.3. Each set of bar graphs represents one group or one score category of a variable. If sex were the grouping variable, one set of bars would refer to males and the other to females. Each set is then drawn in reference to its distribution on the other variable, for example, seriousness of offense.

Table 10.1 provides the joint frequency and percentage distribution for sex, a nominal variable, and for seriousness of offense, an ordinal variable. Percentages should be calculated in the direction of the grouping or independent variable. In this case, sex is the grouping variable, so column percentages have been calculated. Since there are more males than females, a multiple bar graph of percentages rather than frequencies makes the comparison of relative group differences easier. Thus, the Y-axis in Figure 10.2 gives the percent of subjects. The X-axis gives the score intervals for seriousness of offense. For each of its score intervals, one bar is drawn for males and another bar for females. The bar for females has been shaded to make a clear distinction between groups. Note

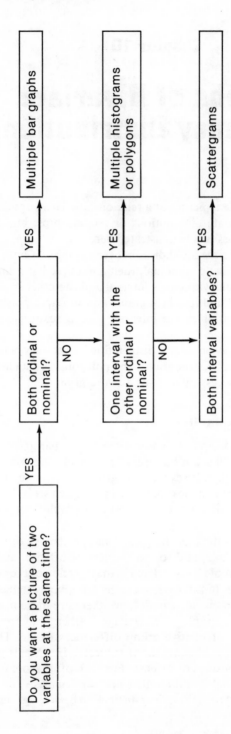

Figure 10.1.   Decision-making chart for graphs of bivariate frequency distribution.

Table 10.1.   Bivariate frequency and percentage (in
parentheses) distribution for sex and seriousness of
offense of juveniles in detention homes (*n* = 70)

Seriousness	Sex	
of offense	Male	Female
Low	10 (25%)	12 (40%)
Moderate	10 (25%)	9 (30%)
High	20 (50%)	9 (30%)
Total	40 (100%)	30 (100%)

that in the upper part of the graph, two boxes are shown to indicate that
the lightly shaded bars represent males and the darkly shaded bars represent females. Lastly, the actual percentage for each bar is placed inside
the bar.

Figure 10.2 highlights the relationship between sex and seriousness of
offense. Females tend to commit relatively more offenses of low seriousness, whereas males tend to commit relatively more offenses of high seriousness. Both females and males commit about the same relative number
of offenses of moderate seriousness. Note the use of the word "relative"
in the above description. This word means that percentages rather than
frequencies are the basis of comparison.

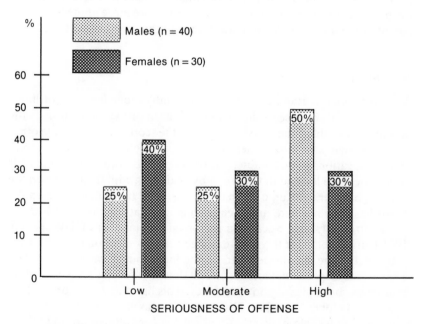

Figure 10.2.   A multiple bar graph for sex and seriousness of offense for juveniles in detention homes (*n* = 70).

10.2   MULTIPLE HISTOGRAMS AND POLYGONS

Since the multiple histograms are drawn like bar graphs with no spaces between bars, only multiple polygons will be discussed here.

Multiple polygons consist of two or more polygons in the same figure, where each polygon represents one group or one score interval on an ordinal or nominal variable. Each polygon is then drawn in reference to its distribution on the interval variable along the *X*-axis. For example, Table 10.2 provides the joint frequency and percent distribution for sex, a nominal variable, and for number of days spent in detention homes, an interval variable. Percentages are calculated in the direction of the grouping variable, which in this case is sex. Because there are more males than females, percentages rather than frequencies are placed on the *Y*-axis of Figure 10.3 to make group comparisons easier. The score intervals for number of days spent in detention homes are put along the *X*-axis. A polygon is then drawn for each group, males and females. The polygon for females consists of a dotted line to distinguish it clearly from the straight-line polygon for males.

Figure 10.3 highlights the relationship between sex and number of days spent in detention homes. Females tend to spend relatively fewer days in detention homes than males. The polygon for females is not only flatter but peaks at 5–10 days, whereas the polygon for males is taller and peaks at 13–14 days. The polygon for females is also more spread out than the polygon for males. All these characteristics are more immediately apparent in the multiple polygon graph than in the distribution table.

10.3   SCATTERGRAMS

A *scattergram* is a graph of two interval variables together on coordinate axes. The *Y*-axis represents the score intervals of one variable (usually the dependent variable), the *X*-axis represents the score intervals of the other variable (usually the independent variable), and points in the graph represent the location of each subject in terms of both variables. Points for each subject are located in the graph by beginning with their score on the *X*-axis, going straight up over that score interval until their score on the *Y*-axis has been reached, and then placing a dot at their intersection.

Figure 10.4 shows a scattergram for age and number of days spent in the detention home for the first 10 subjects in the detention home data list (Table 4.1). Age is considered to be the independent variable and days spent is the dependent variable. Note that the subject's identification number has been placed above each dot. This has only been done to help locate points for each subject. However, they should not ordinarily be put in a scattergram. Note also that a circle has been drawn around all the points. Again, this is usually not done. It is done here only to emphasize

Table 10.2.   Bivariate frequency and percent (in parentheses) distribution for sex and number of days spent in detention homes ($n = 70$)

Days spent	Sex	
	Male	Female
1–2	0 (0%)	2 (7%)
3–4	0 (0%)	5 (17%)
5–6	6 (15%)	6 (20%)
7–8	6 (15%)	6 (20%)
9–10	6 (15%)	6 (20%)
11–12	10 (25%)	3 (10%)
13–14	12 (30%)	2 (7%)
Total	40 (100%)	30 (101%)[a]

[a]Total of 101 percent is due to rounding error.

the relationship between the two variables. The positive increase of the circle from left to right indicates that age and days spent tend to increase together and thus have a positive relationship. The following chapter will make use of scattergrams to illustrate different types of relationships between variables.

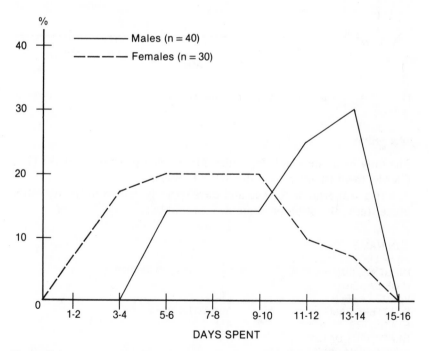

Figure 10.3.   A multiple polygon of sex and number of days spent in detention homes ($n = 70$).

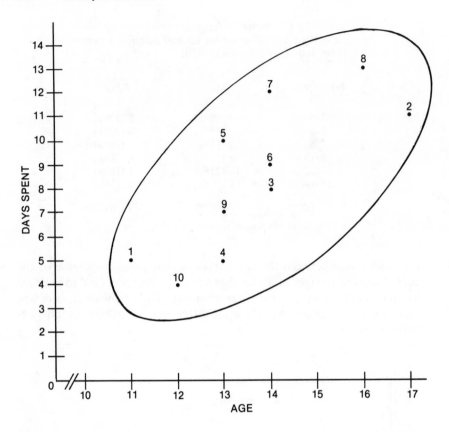

Figure 10.4. A scattergram of age and days spent for the first 10 girls in the detention home.

10.4 SPSS

The only bivariate graph for which SPSS has a program is SCATTER-GRAM (read Chapter 10 in the *SPSS Primer*).

If a scattergram for age and days spent is desired in the detention home study, the SPSS program deck would be:

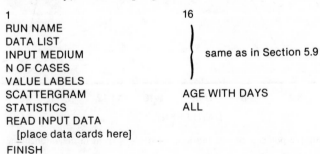

```
1                              16
RUN NAME
DATA LIST
INPUT MEDIUM                   } same as in Section 5.9
N OF CASES
VALUE LABELS
SCATTERGRAM                    AGE WITH DAYS
STATISTICS                     ALL
READ INPUT DATA
   [place data cards here]
FINISH
```

The SCATTERGRAM card indicates the variables in the scattergram. Age will be along the Y-axis, and days will be along the X-axis. In SPSS, the variable on the left of WITH will always be placed on the Y-axis, and the variable on the right of WITH will always be placed on the X-axis. Dots for each subject will be inside the axes to represent the subject's scores on both variables. If there is more than one subject in the same location, SPSS will print the number of subjects instead of a dot.

An example of an SPSS scattergram is not given here because it would take too much space. However, you should be sufficiently comfortable with SPSS to make your own scattergram.

Chapter 11

Measures of
Bivariate Relationship

So far, the relationship between two variables has been assessed by percentage differences and graphs. Although these techniques indicate the existence and direction of a relationship, their disadvantages include: 1) relatively lengthy descriptions, 2) no standard measure of strength, and 3) no method to compare two or more relationships. *Measures of relationship,* however, provide a single number or index that summarizes the existence and strength if not the direction and type of a relationship and allows for comparisons of relationships. Thus, measures of relationship are often the preferred summary method of description.

All measures of relationship use *fixed limits* to describe a relationship. A one (1.00) indicates a perfect relationship and a zero indicates no relationship. As the index number increases from zero to 1, the relationship is said to increase in strength. When ordinal or interval variables are involved, the direction of a relationship is indicated by attaching a sign to the index number. If the index number is between -1.00 and zero, this indicates a negative relationship, where -1.00 is a perfect negative relationship. If the index number is between plus one (1.00) and zero, this indicates a positive relationship, where $+1.00$ is a perfect positive relationship. When the same measure of relationship is used, the strength and direction of relationships between different pairs of variables may be compared, much as using a miles per gallon index allows a comparison of fuel economy across different types of cars.

Measures of relationship also imply *prediction.* When two variables are related, their scores co-vary in a patterned way; thus, scores on either variable may be used to predict scores of the other variable (review Section 9.3). When there is a perfect relationship between variables, the scores of one variable predict the scores of the other variable without any errors. When two variables are not related, knowledge of one variable's scores do not help predict the other variable's scores. Because the concept of prediction makes the interpretation of relationships intuitively easier, the measures of relationship included in this chapter may be interpreted in this way. The specific prediction model employed is called the proportional reduction of error (PRE) model. This model is discussed in Section 11.1.

Measures of relationship, like percentage differences in Section 9.3, may also be based on the distinction between independent and dependent variables. When a measure of relationship indicates the existence and strength of a relationship from only an independent to a dependent variable, it is called an *asymmetric* measure. When a distinction is not made among variables, the measure of relationship indicates the mutual existence and strength between both variables, and it is called a *symmetric* measure. Choice between asymmetric and symmetric measures depends primarily on the purpose of your research, which, of course, will vary from one research project to the next. Generally speaking, however, symmetric measures of bivariate relationship are usually reported in the research literature.

Given the above, choice of an appropriate measure of relationship depends primarily on the *level of measurement* of your variables. Nominal-level measurement only classifies, ordinal-level measurement both classifies and rank-orders information, and interval-level measurement classifies, rank-orders, and establishes intervals or values of an equal size (see Chapter 1). The decision-making chart in Figure 11.1 assumes that both variables are similar in measurement: nominal/nominal, ordinal/ordinal, or interval/interval. If, however, you should have mixed levels of measurement scales, a general rule of thumb is to choose a measure of relationship that fits the variable with the lowest level scale. In the situation of having nominal and ordinal variables, for example, measures of relationship appropriate for nominal scales should be used.

Although level of measurement is very important in choosing an appropriate measure of relationship, it should be pointed out that the measures of relationship for interval variables, Pearson r and r^2, are frequently used for nominal and ordinal variables as well. They may be used for *nominal variables,* for instance, when both variables are dichotomous or only have two score categories. Examples include male/female, white/black, on-welfare/off-welfare, government agency/private agency, and so on. Should a nominal variable have more than two categories, the categories may be collapsed or combined. Religion in the detention home study, for example, has three categories: Protestant, Catholic, and other. But Catholic and other could be combined to make religion a dichotomous nominal variable with the categories Protestant and all others. Separate numbers could then be assigned to each category and the Pearson r or r^2 computed to determine the existence and strength (but obviously, because a nominal variable is used, not the direction) of a relationship. Although it is beyond the scope of this book to demonstrate the legitimacy of using Pearson r or r^2 with any type of dichotomous variable, you should know it is possible.

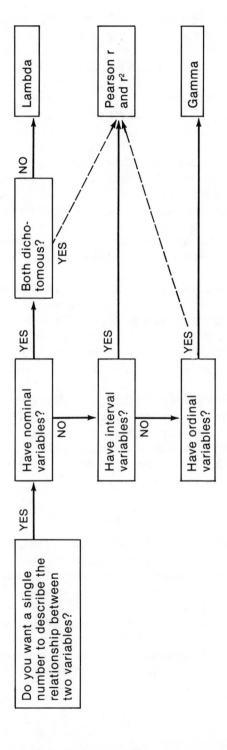

Figure 11.1.   Decision-making chart for measures of bivariate relationship using the PRE model.

Pearson r and r^2 are often used for measuring the relationship between *ordinal variables* as well. This practice is rationalized by arguing that the ordinal variable is theoretically an interval variable that has only been measured, unfortunately, on an ordinal scale. Thus, it should be treated as an interval variable. This practice is more strongly rationalized on empirical grounds by the demonstration that conclusions based on the use of Pearson r and r^2 for ordinal variables, especially as the number of their score intervals increases, seldom differ from conclusions based on the use of ordinal measures of relationship in terms of indicating either the strength or the direction of a relationship.

But why go through such complex arguments to justify this use of Pearson r and r^2 with nominal and ordinal variables? The reason is simple. Most advanced statistical techniques are based on Pearson r and r^2, and it is these techniques which most researchers want to use. Once you learn these techniques, you may want to use such arguments too, but for now you should at least understand the importance of learning about Pearson r and r^2.

Before doing so, however, it is important to understand that calculation of measures of relationship may only be done on pairs of variables for the same subjects. Scores on an empathy test and on an altruism test may only be related if the same subjects have taken both tests. Moreover, the measure of relationship that results may only be used to describe the relationship within that group of subjects. Part III, however, will discuss how to generalize beyond the specific sample.

Finally, throughout this chapter, age and days spent in the detention home study are the variables used to calculate the different measures of relationship. This means that, although both are interval variables, they will also be treated as nominal and ordinal variables. You may wish to review Table 9.6 to get a sense of the variety of measures of relationship that are calculated for you in SPSS.

11.1 PROPORTIONAL REDUCTION OF ERROR

Prediction is a major concern of social science. Thus, to aid interpretation, many measures of association are based on the concept of prediction, or how well one variable predicts another. A perfect relationship is said to exist if all errors of prediction are accounted for. No relationship exists if no errors of prediction are accounted for. The more error is reduced or accounted for, the stronger the relationship is said to be.

The strength of a relationship may therefore be assessed by calculating the proportional reduction in error when using one variable to predict the other. *Proportional reduction of error* (PRE) is defined as:

$$PRE = \frac{B-A}{B}$$

where

  B = the original number of *errors* made in predicting scores of a dependent variable alone *without* using an independent variable as a predictor
  A = the new number of errors made in predicting scores of a dependent variable *with* using an independent variable as a predictor.

It is helpful to think of error by rule B as the amount of error one would make using a measure of central tendency (a summary measure indicating the typical score in a distribution) to predict every score in the distribution. If the central tendency measure is the same as every score, error B is 0. The more the central tendency measure differs from scores in the distribution, the greater error B is. Error B, then, is like a measure of variability (a summary measure of scatter of scores from a central tendency measure). Likewise, one can think of error by rule A as the amount of error one makes in using scores of an independent variable to predict scores of a dependent variable. If this new number of errors is smaller than the original number of errors, then knowledge of the independent variable does help predict; that is, it reduces the original errors of prediction. If the new number of errors is the same as the original number of errors, then knowledge of the independent variable is no more helpful in predicting the dependent variable than was the dependent variable's measure of central tendency.

Thus, proportional reduction in error may be written as:

$$PRE = \frac{\text{original error} - \text{new error}}{\text{original error}}$$

or:

$$PRE = \frac{\text{amount of error reduced}}{\text{original error}}.$$

The definition of PRE, then, simply results in a ratio (see Section 5.6) that indicates the proportion of error reduced when an independent variable is used to predict a dependent variable rather than only using the dependent variable itself. Other synonymous words used for "reduced" are "eliminated," "accounted for," or "explained."

Many measures of relationship are based on such a PRE model. In doing so, two kinds of rules are used: 1) rules that allow the prediction of the dependent variable with an independent variable, or *error by rule B,* and 2) rules that allow the prediction of the dependent variable with an in-

dependent variable, or *error by rule A*. Such measures of relationship ar
generally defined as:

$$\text{measure of relationship} = \frac{\text{error by rule B} - \text{error by rule A}}{\text{error by rule B}}$$

If all error of prediction for a dependent variable can be accounted
for (reduced, eliminated, or explained) by using an independent variable
to predict the dependent variable, a measure of relationship will equal
1.00. If no error of prediction for the dependent variable can be accounted for, a measure of relationship will equal zero. Values between
these fixed limits indicate the proportion of reduced error. For example, if
the original error (*B*) were 20, and use of three different independent variables result in new errors (*A*) of 0, 10, and 20, then the three PRE measures would be:

$$\frac{20-0}{20} = \frac{20}{20} = 1.00$$

$$\frac{20-10}{20} = \frac{10}{20} = 50$$

$$\frac{20-20}{20} = \frac{0}{20} = 0.$$

In the first case, use of the independent variable resulted in no errors of
prediction. It accounted for all original errors of prediction, and thus the
PRE measure equals 1.0 for perfect prediction. In the second case, use of
the independent variable reduced half of the errors of prediction, and thus
the PRE measure equals .50. In the last case, the independent variable did
not account for any errors of prediction, and thus the PRE measure
equals zero.

Each measure of relationship that uses this PRE model differs only in
how error is defined. The following sections discuss such differences.

It should also be noted that, as defined, PRE measures only reflect
the strength and not the direction of a relationship. Thus, use of the PRE
model only helps to interpret the meaning of a relationship in terms of
predictive strength. However, the calculation of PRE measures does not
result in the demonstration of the direction of a relationship.

11.2 LAMBDA

Lambda, or Guttman's coefficient of predictability, is an appropriate
choice for measuring the relationship between two nominal variables or
any two variables treated as nominal variables. Since lambda is designed

for nominal variables, it cannot logically show the direction of a relationship. Its index will only vary, therefore, from zero (no relationship) to 1 (perfect relationship). The symbol for a symmetric lambda is the Greek letter λ. The symbol for an asymmetric lambda is λ_{YX}, where the first subscript (Y here) refers to the dependent variable and the second subscript (X here) refers to the independent variable.

Asymmetric lambda reflects only the predictive strength of relationship from an independent variable to a dependent variable. Thus, in the relationship between age and days spent, age would be the independent variable or X, and days spent would be the dependent variable or Y, as shown in Table 11.1.

Calculation of asymmetric lambda uses the mode to define error. The mode is the most frequently occurring score interval in a univariate distribution and is the appropriate measure of central tendency for nominal variables. Use of this measure of central tendency will result in the smallest number of errors in prediction.

Error by rule B is defined in reference to the mode for dependent variable, age. The modal category is 1–7 days, or 1 week, and contains 16 out of the 30 girls in the detention home. If one guessed the mode as the typical score for all girls in the home, one would make 14 errors (30 − 16 = 14). *Error by rule B is thus defined as the number of nonmodal subjects in the dependent variable distribution.*

Error by rule A is defined in reference to the mode of the dependent variable for each score interval of the independent variable. The modal category of 10–13 years of age is 1–7 days, because it contains 11 out of the 15 girls. If one guessed 1–7 days as the typical score for all girls 10–13 years of age one would make four errors (15 − 11 = 4). The modal category of 14–17 years of age is 8–14 days, because it contains 10 out of the 15 girls. If one guessed 8–14 days as the typical score for all girls 14–17 years of age, one would make five errors (15 − 10 = 5). *Error by rule A is defined as the sum of nonmodal subjects in each of the score intervals of the independent variable.* Thus, when age is used to predict days spent, there are nine (4 + 5 = 9) errors of prediction.

Table 11.1. Contingency table of age and days spent in the detention home ($n = 30$)

Days spent	Age		Total
	10–13	14–17	
1–7	11	5	16
8–14	4	10	14
Total	15	15	30

Asymmetric lambda may now be calculated as:

$$\lambda_{YX} = \frac{\text{error by rule B} - \text{error by rule A}}{\text{error by rule B}}$$

$$= \frac{14-9}{14}$$

$$= .36.$$

Thus, 36 percent of the errors of prediction concerning days spent in the detention home are accounted for by taking age into account. The strength of relationship between the two variables reflects the relative improvement (36 percent) in prediction attainable by using age as the independent variable.

If days spent were to become the independent variable and age the dependent variable, then:

$$\lambda_{XY} = \frac{15 - (5+4)}{15}$$

$$= .40.$$

Thus, 40 percent of the errors of prediction concerning age are eliminated by taking days spent in the detention home into account. Although this lambda makes no sense logically, it does highlight the fact that designation of independent and dependent variables is important in calculating asymmetric lambda.

Symmetric lambda is a kind of average of both asymmetric lambdas (λ_{YX} and λ_{XY}):

$$\lambda = \frac{\Sigma(\text{error by rule B} - \text{error by rule A})}{\Sigma(\text{error by rule B})}$$

$$= \frac{(14-9)+(15-9)}{14+15}$$

$$= .38.$$

Thus, 38 percent of the errors of mutual prediction are eliminated or accounted for. As such, there is an overall 38 percent improvement when predictions are made each way.

Although there are other measures of relationship for nominal variables (for example, *phi,* the *contingency coefficient,* and the *uncertainty coefficient)* that SPSS calculates (see Table 9.6), they are not PRE measures and thus are more difficult to interpret. Consequently, one seldom sees these measures reported in the literature.

Lastly, even though age and days spent were treated as nominal variable in this discussion, the very fact that they are both dichotomous variables as well means that the interval-level measures of relationship (Pearson r and r^2) could have been used instead of lambda. This will be discussed further in Section 11.4.

11.3 GAMMA

Gamma is an appropriate choice for measuring the relationship between two ordinal variables or for the mix of an ordinal and interval variable. Since gamma applies to at least ordinal-level variables, it shows not only the strength but also the direction of a relationship. Gamma is a symmetric measure of relationship. The fixed limits of gamma are, therefore, -1.00 and 1.00. The symbol for gamma is the Greek letter γ.

The unit of analysis for gamma is a *pair* rather than a single subject. If the relative rank order between a pair of individuals is known for one variable, how accurately does this knowledge predict relative rank order for the same pair on the other variable? Would this prediction be any better than random guessing of rank order? Gamma indicates the superiority of predicting rank order within a pair, given previous knowledge of rank order on another variable, relative to random guessing.

The first step to understanding gamma is learning how to calculate relative rank order for pairs of subjects. Gamma uses two types of rank order among pairs. One type is *those pairs that show the same rank order (positive relationships) for individuals on both variables.* The symbol for such pairs is N_s. To calculate N_s, multiply the frequency of each cell in a contingency table by the total of all frequencies in the cells below it *and* to its right, and then add all the products together. The other type is *those pairs that show an inverse rank order (negative relationships) for individuals on both variables.* The symbol for such pairs is N_d. To calculate N_d, multiply the frequency of each cell in a contingency table by the total of all frequencies in the cells below it *and* to its left, and then add all the products together.

To illustrate these calculations, refer to Table 11.1, which shows the bivariate frequency distribution for age and days spent. The number of pairs that show the same rank order is:

$$N_s = 11 \times 10$$
$$= 110.$$

The number of pairs that show an inverse rank order is:

$$N_d = 5 \times 4$$
$$= 20.$$

Because N_s is greater than N_d, you should suspect a positive relationship between age and days spent. Indeed, the *direction of gamma* is determined by which one is greater. If N_s is greater than N_d, the relationship is *positive*. If N_d is greater than N_s, the relationship is *negative*.

These two types of pairs are also used to define error. *Error by rule A* is the amount of error made in predicting rank order within pairs for one variable on the basis of rank order within pairs on another variable. In comparing N_s and N_d, error by rule A is the smallest one or minimum (N_s or N_d). N_d is the smallest one in the above example, so error by rule A would be 20. *Error by rule B* is the amount of error made in randomly guessing rank order within pairs for one variable without knowledge of rank order on another variable. Given the total number of pairs ($N_s + N_d$), random guessing should result in a guess of 50 percent of pairs being of the same rank order and of 50 percent of pairs being of inverse rank order. (The use of random guessing here might be puzzling to understand at first, but recall that the median is the scale point which divides subjects into half falling below and half falling above that point.) Error by rule B thus equals $.5(N_s + N_d)$. In the above example, error by rule B is $.5(110 + 20) = 65$.

The PRE formula for gamma is:

$$= \frac{\text{error by rule B} - \text{error by rule A}}{\text{error by rule B}}$$

$$= \frac{.5(N_s + N_d) - \text{minimum}(N_s \text{ or } N_d)}{.5(N_s + N_d)}.$$

The gamma for age and days spent would therefore be:

$$= \frac{.5(110 + 20) - (20)}{.5(110 + 20)}$$

$$= \frac{45}{65}$$

$$= .69.$$

There is a 69 percent proportional reduction in error by using each variable to predict the other. Because .69 is much closer to 1.00 than to zero, the relationship between age and days spent is quite strong.

The relationship between age and days spent is also positive, because N_s is greater than N_d. If you want the calculation of gamma to include the direction of relationship as well as its strength, then the following formula should be used:

$$= \frac{N_s - N_d}{N_s + N_d}$$

$$= \frac{110 - 20}{110 + 20}$$

$$= +.69.$$

This formula simplifies the calculation of gamma but does not show how it is a PRE measure.

There are other PRE measures of relationship for ordinal variables as well: *Kendall's tau b, Kendall's tau c,* and *Somer's D.* Their calculation is somewhat more complex than for gamma, and generally their value will be lower than for gamma. Their interpretation is also not as straightforward. Moreover, Kendall's tau b is only appropriate for a table which has the same number of rows and columns. Gamma, then, is the simplest measure to use.

Spearman's rho (ρ), which is not shown in Table 9.6, and which is not a PRE measure, is yet another measure of relationship for ordinal variables. It is not used very often, however, because its interpretation is more difficult than for a PRE measure, because it does not lend itself easily to more complex statistical analyses, and because Pearson r may be used in its place without overestimating the strength of relationship.

11.4   PEARSON r AND r^2

Pearson r is a measure of relationship for interval variables which, when squared (r^2), is a PRE measure. Both are symmetric measures of relationship. Pearson r shows the strength and direction of a relationship, and thus its fixed limits range from -1.00 for a perfect negative relationship to $+1.00$ for a perfect positive relationship. Pearson $r,$ however, is not a PRE measure of relationship. The square of Pearson $r,$ or $r^2,$ shows the strength of a relationship in a PRE sense but not its direction, and thus only ranges from zero to $+1.00$. However, r^2 is easier to interpret than Pearson r. Each, therefore, has its own advantage.

The symbols of r and r^2 indicate that the Pearson correlation is based on subjects' scores in a sample, and thus may either describe the extent of relationship in the sample or estimate the unknown correlation in the population. The symbols ρ (the Greek letter rho) and ρ^2 indicate that the Pearson correlation is based on subjects' scores in the total population. However, unless the distinction between a sample value and a population value is important, r and r^2 are generally used.

Both Pearson r and r^2 are based on the assumption of *linearity.* This assumption means that the relationship between two variables is assumed to be linear. As the scores for one variable increase (decrease), scores on the other variable should also tend to increase (decrease). A straight line in a scattergram of data points or joint scores, then, best represents a linear pattern. Figure 11.2 shows several examples. Example A shows no relationship between variables. Example B shows two variables that have a curvilinear rather than a linear relationship. Use of Pearson r or r^2 in this situation would underestimate the actual extent of relationship be-

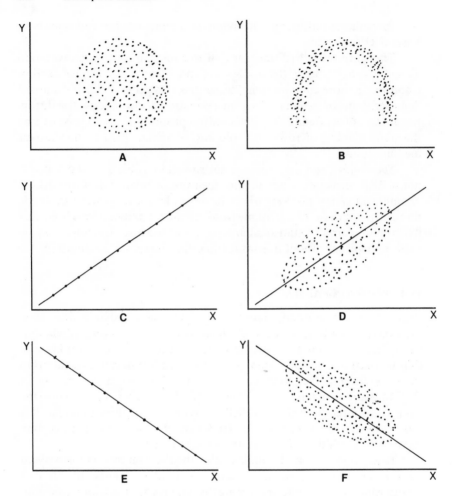

Figure 11.2.   Types of relationships.

tween the two variables. The remaining examples show different types and degrees of linear relationships. Examples C and D, in which the straight line slopes upward to the right, are positive relationships. When all data points fall on the straight line, as in C, then a perfect positive relationship exists. As the data points become more scattered, as in D, the degree of relationship becomes smaller and will become zero if the scatter becomes circular as in example A. Examples E and F, in which the straight line slopes downward to the right, are negative relationships. Example E is a perfect negative relationship, whereas example F is a less strongly negative relationship. Section 11.7 discusses this linear assumption further in terms of simple linear regression.

The formula for defining Pearson r for any two variables X and Y is:

$$r = \frac{\Sigma z_X z_Y}{n}.$$

where

z_X = standard scores for the X variable
z_Y = standard scores for the Y variable
$z_X z_Y$ = the sum of crossproducts of standard scores for each subject on both variables
n = total number of subjects.

This is called the Pearson product-moment correlation coefficient, or simply Pearson r. *Pearson r, then, is the average of crossproducts for two standardized variables.*

To calculate Pearson r, each variable must be put into standard scores (review Section 8.7). The standard scores of both variables for each subject must then be multiplied together to get a crossproduct, the crossproducts must then be summed for all subjects, and lastly the sum of crossproducts should be divided by the number of subjects to get the average crossproduct. Table 11.2 does this for age and days spent in the detention home study. The first column is simply the subject's identification number, the second column is the age of each subject (the X variable), and the third column is the number of days spent in the detention home for each subject (the Y variable). The means and standard deviations for each variable are at the bottom of the columns. The third and fourth columns are the standard scores for each variable, where

$$z_X = \frac{X - \overline{X}}{\sigma_{x^2}}$$

$$z_Y = \frac{Y - \overline{Y}}{\sigma_{y^2}}.$$

For example, the age of the first subject is 11 and the number of days she spent in the detention home is 5. The standard score for her age is:

$$z_X = \frac{11 - 13.5}{1.893} = -1.321.$$

In other words, she is 1.321 standard deviations below the mean for age. The standard score for her days spent is:

$$z_Y = \frac{5 - 7.167}{3.239} = -.669.$$

That is, she is .681 standard deviations below the mean for number of days spent in the detention home. In Table 11.2 the standard deviations

Table 11.2. Calculations for Pearson r and simple linear regression

(1)	(2)	(3)	(4)	(5)	(6)	(7)	(8)
	Age	Days	Deviation scores			Z scores	
ID	X^i	Y^i	$(X^i - X)$	$(Y^i - Y)$	Z_x	Z_y	$Z_x Z_y$
1	11	5	−2.500	−2.167	−1.321	−.680	.898
2	17	11	3.5	3.833	1.843	1.204	.226
3	14	8	.5	.833	.264	.262	.069
4	13	5	−.5	−2.167	−.264	−.680	.180
5	13	10	−.5	2.833	−.264	.890	−.235
6	14	9	.5	1.833	.264	.576	.152
7	14	12	.5	4.833	.264	1.518	.401
8	16	13	2.5	5.833	1.321	1.832	2.419
9	13	7	−.5	−.167	−.264	−.052	.014
10	12	4	−1.5	−3.167	−.793	−.994	.788
11	15	3	1.5	−4.167	.793	−1.309	−1.037
12	10	8	−3.5	.833	−1.843	.262	−.484
13	15	9	1.5	1.833	.793	.576	.456
14	14	5	.5	−2.167	.264	−.680	−.180
15	11	1	−2.5	−6.167	−1.321	−1.937	2.556
16	12	6	−1.5	−1.167	−.793	−.366	.290
17	14	9	.5	1.833	.264	.576	.152
18	15	2	1.5	−5.167	.793	−1.623	−1.286
19	16	14	2.5	6.833	1.321	2.146	2.834
20	13	9	−.5	1.833	−.264	.576	−.152
21	15	4	1.5	−3.167	.793	−.994	−.788
22	14	11	.5	3.833	.264	1.204	.318
23	17	8	3.5	.833	1.849	.262	.484
24	16	7	2.5	−.167	1.321	−.052	−.069
25	12	4	−1.5	−3.167	−.793	−.994	.788
26	13	7	−.5	−.167	−.264	−.052	.014
27	13	9	−.5	1.833	−.264	.576	−.152
28	12	6	−1.5	−1.167	−.793	−.366	.290
29	11	5	−2.5	−2.167	−1.321	−.680	.899
30	10	4	−3.5	−3.167	−1.849	−.994	1.840
Mean	13.500	7.167			.000	.000	
St. Dev.	1.893	3.184			1.000	1.000	
Sum			.0	.000			13.688

for age and days are calculated as population values (σ) rather than as estimates of the population values (s), and these descriptive population values (σ) are in the denominator of each standard score. This procedure is generally followed when calculating the sample correlation coefficient, because little bias is introduced. The calculation of Pearson r, therefore, is the same for population and sample data. Note also in Table 11.2 that, as for all standard scores, the mean and standard deviation of z scores for both variables are zero and 1. The eighth column is the crossproduct of standard scores for both variables on each subject. For example, the crossproduct for the first girl is .87 = (−1.299) (−.669). The sum of the

Table 11.2 (cont.)

(9)	(10)	(11)	(12)	(13)	(14)
			Predicted		Squared
Squared deviations		Crossproduct	score	Error	error
$(X_i - X)^2$	$(Y_i - Y)^2$	$(X_i - X)(Y_i - Y)$	Y_i	$(Y_i - Y_i)$	$(Y_i - Y_i)^2$
6.25	4.695	5.417	5.248	−.248	.062
12.25	14.694	13.417	9.853	1.147	1.316
.25	.694	.417	7.550	.450	.202
.25	4.695	1.083	6.783	−1.783	3.179
.25	8.028	−1.417	6.783	3.217	10.349
.25	3.361	.917	7.550	1.450	2.101
.25	23.361	2.417	7.550	4.450	19.799
6.25	34.027	14.583	9.085	3.915	15.325
.25	.028	.083	6.783	.217	.047
2.25	10.028	4.750	6.016	−2.016	4.062
2.25	17.361	−6.250	8.318	−5.318	28.279
12.25	.694	−2.917	4.481	3.519	12.386
2.25	3.361	2.750	8.318	.682	.465
.25	4.695	−1.083	7.550	−2.550	6.504
6.25	38.028	15.417	5.248	−4.248	18.046
2.25	1.361	1.750	6.016	−.016	.000
.25	3.361	.917	7.550	1.450	2.101
2.25	26.695	−7.750	8.318	−6.318	39.915
6.25	46.694	17.083	9.085	4.915	24.155
.25	3.361	−.917	6.783	2.217	4.915
2.25	10.028	−4.750	8.318	−4.318	18.644
.25	14.694	1.917	7.550	3.450	11.900
12.25	.694	2.917	9.853	−1.853	3.433
6.25	.028	−.417	9.085	−2.085	4.348
2.25	10.028	4.750	6.016	−2.016	4.062
.25	.028	.083	6.783	.217	.047
.25	3.361	−.917	6.783	2.217	4.915
2.25	1.361	1.750	6.016	−.016	.000
6.25	4.695	5.417	5.248	−.248	.062
12.25	10.028	11.083	4.481	−.48¢	.231
			7.167		
107.500	304.167	82.500		.000	240.853

crossproducts in the eighth column is the numerator of the Pearson r formula, which must be divided by 30 ($n = 30$) to get:

$$r = \frac{13.688}{30}$$

$$r = .46.$$

Knowing how to calculate Pearson r, however, does not indicate how to interpret it. Pearson r may be interpreted in two ways. The *first* method of interpretation uses Pearson r to indicate the direction and strength of a linear relationship. A negative sign shows a negative relationship, and a positive sign shows a positive relationship. Thus, for age and days spent, the positive sign of r indicates that the scores for each variable tend to increase together. The strength of the linear relationship is assessed by the rule of thumb presented in Table 11.3. The closer r comes to 1.00 or -1.00, the stronger the linear relationship is said to be. The r for age and days spent could thus be described as a moderate positive correlation. These adjectives, however, only approximate the strength of relationship, and indeed the cutting points for r along each descriptive adjective in Table 11.3 vary somewhat among social scientists. Moreover, the descriptive range for r represents an ordinal scale. It is incorrect, therefore, to say that $r = .60$ is twice as large as $r = .30$ or that an increase from $r = .10$ to $r = .20$ has the same meaning as an increase from $r = .30$ to $r = .40$. These problems make r relatively difficult to interpret.

A *second* way to interpret Pearson r is to square it. If the Pearson r is squared (r^2), the strength of the linear relationship may be interpreted by the proportional reduction of error model. The reason why r^2 is a PRE measure of relationship is provided in more detail in the discussion of simple linear regression at the end of this chapter. For the meantime, the r^2 for age and days spent is:

$$r^2 = (.46)^2 = .21.$$

That is, 21 percent of the error is accounted for in predicting scores on one variable with knowledge of its linear relationship to another variable. Because almost 80 percent of the error remains to be explained, the correlation of $r = .46$ may only be said to be of moderate strength. Thus r^2 offers a foundation for the interpretation of Pearson r as given in Table 11.3.

At the beginning of this chapter, it was mentioned that Pearson r and r^2 could be used legitimately as measures of relationship for dichotomous nominal variables. When this is done, however, the direction of relationship indicated by Pearson r will have no meaning because the category numbers are assigned arbitrarily. (The sign of Pearson r will only reflect how the group, which was assigned a "1", relates to the dependent vari-

Table 11.3.   Rule-of-thumb interpretation for the size of Pearson r

Value of r	Descriptive adjective	Value of r^2
.00 to .20 (.00 to −.20)	Very low	.00 to .04
.20 to .40 (−.20 to −.40)	Low	.04 to .16
.40 to .60 (−.40 to −.60)	Moderate	.16 to .36
.60 to .80 (−.60 to −.80)	High	.36 to .64
.80 to 1.00 (−.80 to −1.00)	Very high	.64 to 1.00

able.) Otherwise, Pearson r and r^2 may be interpreted as usual. However, do not expect either Pearson r or r^2 to be of the same size as lambda. Each measure of relationship is different and, thus, comparisons may only be made when the same measure is used.

Before concluding the discussion of Pearson r, several observations should be stated. First, *the size of Pearson r is not affected by sample size.* Smaller correlation coefficients are not caused by small sample sizes. Increasing the size of your sample, therefore, will not necessarily result in a larger correlation coefficient. Second, *the size of Pearson r is affected by a restriction of range in scores.* If there is no variability of scores in a variable, as measured by the variance or standard deviation, then logically there is no way that changes in another variable can predict changes in a variable that does not change. Generally, as the variability of either variable becomes smaller, so does the value of the correlation coefficient. This may be demonstrated by the Pearson r for age and days spent, which is $r = .46$ when the full range of scores is used. When their scores are dichotomized, however, as in Table 11.1, the Pearson r is .40 (see Table 9.6 for a confirmation of this Pearson r). The restriction in range for both variables results in a smaller Pearson r. A similar argument is used when a Pearson r is calculated for ordinal variables. The relatively smaller number of score intervals for ordinal variables is presumed to make Pearson r an underestimate of the real strength of their relationship. As the number of score intervals for ordinal variables increases, therefore, the more likely is Pearson r to reflect their relationship. Finally, *the size of Pearson r is affected by outliers*, or the presence of infrequent extreme scores. A moderate correlation may be raised to a high correlation if several extreme scores allow a better straight line to be drawn. Conversely, a moderate correlation may be lowered to near zero if several extreme scores fall far away from the straight line that would ordinarily be drawn through the cluster of score points. A look at a scattergram will indicate whether outliers have influenced the size of Pearson r.

11.5   THE CORRELATION MATRIX

Because it is often informative to compare relationships between pairs of different variables, a correlation matrix provides a visual way of listing all correlation coefficients. Although a matrix may consist of any measure of relationship, each matrix should only use the same measure, because otherwise the comparisons would be meaningless.

The most frequent type of correlation matrix is the one for Pearson r. Table 11.4 shows an example of such a matrix for all variables in the detention home study. Note that Pearson r is used for the ordinal variable "seriousness of offense," and for the nominal variable "religion." The

Table 11.4.   Pearson r correlation matrix for all variables in the detention home study ($n = 30$)

	Age	Religion	Days	Offense
Age	1.00	−.23	.46	.62
Religion		1.00	−.14	−.04
Days			1.00	.40
Offense				1.00

ordinal variable, therefore, is being treated as an interval variable. The nominal variable was made into a dichotomous variable by arbitrarily assigning a score of one to the category "Protestant" and a score of zero to the categories "Catholic" and "other." The nominal variable may then be treated as an interval variable as well.

The rows and columns of a correlation matrix refer to the variables. The intersection between each row and column variables shows their Pearson r coefficient. To find the Pearson r for age and days spent, for example, find the age row and read over to the days spent column. Their intersection is at $r = .46$.

As in this matrix, correlations are usually put only in the top part of the table. This is done not only because it communicates all the observed correlations but also because it avoids the necessity of writing them all again in the bottom part of the table (remember that Pearson r is a symmetric measure of association). Note as well that the diagonal of the matrix is a series of 1.00s. They may be omitted if desired, because they only indicate the obvious—a variable is always perfectly correlated with itself.

What story may now be told about the variables in the detention home study? Their measures of central tendency, variability, and relationship are known, so certainly something can be said, but what? Think about it and then read the next chapter.

11.6   SPSS

Measures of relationship for nominal and ordinal variables, including lambda and gamma, are calculated by SPSS in the CROSSTABS program. Section 9.5 and Table 9.6 summarize the SPSS procedures and printout.

Pearson r is usually done by means of the PEARSON CORR program, discussed in Section 18 of the *SPSS Manual*. In the *SPSS Primer*, however, Pearson r must be done by using the SCATTERGRAM procedure. Try it both ways. The results will be exactly the same.

I used the following set of cards to get the Pearson correlation coefficients in Table 11.4:

```
1                                16
RUN NAME
DATA LIST                        ⎫
INPUT MEDIUM                     ⎬    same as in Section 5.9
N OF CASES                       ⎭
RECODE                           RELIGION (1 = 1) (2,3 = 0)/
VALUE LABELS                     RELIGION (1) PROTESTANT
                                          (0) ALL OTHERS/
PEARSON CORR                     ALL
READ INPUT DATA
   [place data cards here]
FINISH
```

The RECODE card was used to recode "religion" into a two-category nominal variable. The PEARSON CORR card tells the computer to calculate Pearson r among all variables.

The printout, as shown in Table 11.5, provides a correlation matrix. The top number is the Pearson r. The number underneath it in parentheses is the number of subjects used to calculate Pearson r. This number will always be equal to your sample size, unless you have missing data for one or both variables. The last number is the significance level for Pearson r, a subject to be discussed in Part III.

Table 11.5.   SPSS printout of a Pearson r correlation matrix

	SUBJECT	AGE	RELIGION	DAYS	OFFENSE
SUBJECT	1.0000	− 0.1617	0.1964	− 0.2038	− 0.0811
	(0)	(30)	(30)	(30)	(30)
	S = .001	S = .197	S = .149	S = .140	S = .335
AGE	− 0.1617	1.0000	− 0.2289	0.4562	0.6254
	(30)	(0)	(30)	(30)	(30)
	S = .197	S = .001	S = .112	S = .006	S = .001
RELIGION	0.1964	− 0.2289	1.0000	− 0.1361	− 0.0401
	(30)	(30)	(0)	(30)	(30)
	S = .149	S = .112	S = .001	S = .237	S = .417
DAYS	− 0.2038	0.4562	− 0.1361	1.0000	0.3970
	(30)	(30)	(30)	(0)	(30)
	S = .140	S = .006	S = .237	S = .001	S = .015
OFFENSE	− 0.0811	0.6254	− 0.0401	0.3970	1.0000
	(30)	(30)	(30)	(30)	(0)
	S = .335	S = .001	S = .417	S = .015	S = .001

11.7 SIMPLE LINEAR REGRESSION

The Pearson correlation coefficient may also be described in terms of simple linear regression. Indeed, the terms of correlation, regression, and prediction are often used interchangeably. When doing so, however, it is assumed that only interval-level variables are being used.

Simple linear regression refers to the situation in which one independent variable is used to predict a dependent variable, whereas multiple regression refers to the situation in which two or more independent variables are involved. This book only describes the one-variable situation.

The basis of simple linear regression is the straight line for which the equation is:

$$Y_i = a + bX_i$$

where

Y_i = the symbol for scores of the dependent variable

a = a constant called the *Y intercept*. It reflects the value of Y when the value of the X variable is zero, or, in other words, it is the point at which the straight line intersects the Y-axis.

b = the *slope* of the straight line. Assuming that all (X, Y) points fall on the straight line, it is calculated as:

$$\frac{Y_2 - Y_1}{X_2 - X_1}.$$

That is, the slope is the rate at which Y changes with a change in X. Its value may be found by selecting any two values of Y and the corresponding two values for X. A positive slope means that X and Y values increase together (example C in Figure 11.2); a negative slope means the X values increase or decrease as Y values decrease or increase (example E in Figure 11.2)

X_i = scores on the independent variable that are used to predict the dependent variable.

However, it is most unlikely in the real world that one set of scores (X_i) will perfectly predict another set of scores (Y_i) in terms of all scores falling directly on a straight line, as shown in examples C and E in Figure 11.2. Instead, it is much more likely that scores will cluster to some extent around a straight line, as shown in examples D and F in Figure 11.2. This latter straight line is called a regression line. A *regression line* is the hypothetical straight line that results from assuming that a linear relationship exists between two variables and that results from using a regression equation. A *regression equation* is the equation used to define the line in such a way that the least amount of error is made in predicting dependent variable scores from independent variable scores. The purpose of regression, therefore, is to minimize errors of prediction around the regression line:

$$\text{error} = Y_{i_t} - Y_i'$$

where

$Y_i' =$ a predicted Y_i which is always on the straight regression line and is called "Y prime." The discrepancy between the real or observed value of Y_i and its predicted value (Y_i') is the error of prediction.

Error of prediction is minimized by using a *least-squares criterion* (see Sections 7.5 and 8.4) to define the Y intercept (a) and the slope (b) of a regression line. The resulting prediction equation, or regression equation, defines a regression line as:

$$Y_i' = a + bX_i$$

where

$a = \overline{Y} - b\overline{X}$

Here, $\overline{Y}$ is the mean of the observed scores for the dependent variable Y, and $\overline{X}$ is the mean of the observed scores for the independent variable X

$$b = \frac{\Sigma(X_i - \overline{X})(Y_i - \overline{Y})}{(X_i - \overline{X})^2} = \frac{\Sigma xy}{x^2}.$$

Here, $(X_i - X)$ is the deviation of X scores from the mean of X and $(Y_i - Y)$ is the deviation of Y scores from the mean of Y. The numerator is the sum of crossproducts of these deviation scores for each subject. The sum of crossproducts for all subjects is divided by the sum of squared deviations for each subject on the X variable. The slope constant that results is called the *regression coefficient*.

To illustrate the use of this regression equation, consider days spent in the detention home as the dependent variable and age as the independent variable. Assuming that a linear relationship exists between these variables, then using the information in Table 11.2:

$$b = \frac{82.500}{107.500} = .767$$

where the numerator is the sum of crossproducts from column 6 (found by multiplying columns 4 and 5 for each subject), and the denominator is the sum of squared deviations about the mean from column 2

$a = 7.17 - .767(13.5) = -3.19$

where 7.17 is the mean of days spent, .767 is the slope, and 13.5 is the mean of age.

The slope or regression coefficient indicates that, with a unit change (1 year) in age, there is a .767 change in days spent. The positive sign represents a positive slope for the regression line. The Y intercept, or a, indicates that the regression line intersects the Y axis at -3.19 years.

Given a known slope and Y intercept, it is possible to predict individual scores on days spent with knowledge of age on the basis of the regression equation. For example, the predicted scores for the first three subjects in Table 11.2 are:

$$Y_1 = -3.19 + .767 \ (11) = 5.25 \ \text{days}$$
$$Y_2 = -3.19 + .767 \ (17) = 9.85 \ \text{days}$$
$$Y_3 = -3.19 + .767 \ (14) = 7.55 \ \text{days}.$$

The remaining predicted scores on days spent may be found in column 7 of Table 11.2.

The question of how well knowledge of age predicts days spent is answered by comparing the error or difference between the predicted and observed scores on days spent. For example, the errors made in predicting scores for the first three subjects in Table 11.2.

$$\text{error}_1 = 5 - 5.25 = -25 \ \text{days}$$
$$\text{error}_2 = 11 - 9.85 = 1.15 \ \text{days}$$
$$\text{error}_3 = 8 - 7.55 = .45 \ \text{days}.$$

The remaining errors made for the other subjects may be found in column 8 of Table 11.2. These errors represent deviations of the observed scores from the predicted scores on the regression line $(Y_i - Y_i')$. Just as for the mean (see Section 7.5 on the least-squares property), when these deviations are squared and summed, the resulting sum is a minimum value. Thus, the sum of squared deviations about predicted scores represents a minimum estimate of error of prediction when using age to predict days spent. This sum of squared deviations is called *unexplained variation* in days spent; that is, it is the variation in days spent that cannot be predicted or accounted for by knowledge of a subject's age. If age could predict days spent perfectly, this sum would equal zero and all joint scores (X, Y) for each subject would fall exactly on the regression line. Unexplained variation is also the *error by rule A* in the PRE model:

$$\text{unexplained variation} = \text{error by rule A}$$
$$= \Sigma (Y_i - Y_i')^2.$$

In the present example, error by rule A is the sum of squared deviations about predicted scores in column 9 of Table 11.2 and equals 240.85. Error by rule B is the error obtained in predicting scores on the dependent variable without knowledge of an independent variable. This error is often called the *total variation* of the dependent variable. It is simply the sum of squared deviations about the dependent variable's mean, or the numerator of the variance formula:

$$\text{total variation} = \text{error by rule B}$$
$$= \Sigma(Y_i - \overline{Y})^2.$$

In the present example, error by rule B is the sum of squared deviations about the mean of days spent in column 3 of Table 11.2 and equals 304.17. These two estimates or error may now be used to construct an interval measure of relationship that reflects the proportional reduction in error in using a linear regression equation to predict a dependent variable. The symbol for this measure of relationship is r^2:

$$r^2 = \frac{\text{total variation} - \text{unexplained variation}}{\text{total variation}}$$
$$= \frac{\Sigma(Y_i - \overline{Y})^2 - \Sigma(Y_i - Y_i')^2}{\Sigma(Y_i - \overline{Y})^2}$$
$$= \frac{304.17 - 240.85}{304.17}$$
$$= .21.$$

That is, in the present example, 21 percent of the variation in days spent can be accounted for by knowledge of a subject's age. Moreover, the square root of r^2 is the absolute value (see Section 2.7) of Pearson r:

$$|\text{ Pearson } r | = \sqrt{r^2} = \sqrt{.21} = .46.$$

In sum, then, the linear assumption underlying Pearson r correlation coefficients is demonstrated by the use of simple linear regression. Moreover, by using regression procedures, it should be clearer as to how a squared Pearson r (r^2) may be interpreted as a PRE measure of relationship for two interval-level variables.

Chapter 12

The Research Report

Stories and research reports share common characteristics. Both have a structure. The structure of a story revolves around a beginning, middle, and end, as does the structure of a research report. Both have a plot. A story plot involves themes, characters, and personal relationships. A research report involves conceptual frameworks, hypotheses, and empirical relationships. Both require creative imagination to develop the appropriate structure and plot.

Unlike stories, however, research reports depend on how well empirical data either support the beginning of the story or suggest a new story by the way the story ends. Also unlike stories, the elegance of the report depends not so much on the author's style, but on how well the conceptual framework and empirical findings blend to produce empirically convincing conclusions that may be replicated by others. Thus, the acceptance of the story within a research report must be based, first, on the empirical support for the story, and second, on the ease with which others can replicate the empirical findings which support the story.

To accomplish these goals, the structure of a research report has been somewhat standardized over the years. The general structure of a research report is as follows:

1. Executive summary or abstract
2. Conceptual framework
   a. a statement of the research problem
      i. why this problem is important from a theoretical or conceptual perspective
      ii. why this problem is important from an applied or intervention perspective
   b. a literature review to support the conceptual framework
   c. a statement of the hypotheses to be tested or research questions raised
3. Methodology
   a. a description of the subjects
   b. a description of the research design
   c. a description of the variables used to measure the concepts within the hypotheses or research questions

d. a description of the statistics used to test the hypotheses or research questions
4. Results—a summary of research results and data analysis
5. Discussion and conclusions
    a. a discussion about the results which allows for an interpretation of how well they fit the conceptual framework
    b. a statement of conclusions and their meaning.

Although not all elements of this structure need to be present in a research report, it may be said that the quality of the report improves with their inclusion. The executive summary or abstract section states all the important aspects of the other sections in 150 to 200 words. The conceptual framework and discussion and conclusions sections provide the story. The methodology section makes your research procedures explicit so others may replicate your study to ensure its credibility. The results section provides the empirical support for your discussion and conclusions. Whenever possible, then, this structure should be used to develop the plot of research reports.

How may this structure be used to describe the story in the detention home study? Recall that the detention home study just sort of happened. The variables and data were introduced in Chapter 4 and were used thereafter as examples of various kinds of descriptive statistics. No conceptual framework was offered at all. Now there is the request to tell a story on the spot. Let me briefly explain how to get out of this tight predicament.

A conceptual framework provides a research problem, discusses the important concepts, and suggests how these concepts are related to each other by means of hypotheses. Usually the report begins with a theory or model that leads to specific and testable hypotheses. However, when lacking a theory or model, use of the time sequence principle is often helpful. The time sequence principle, in a sense, is going through the back door to get a conceptual framework. Here is how it works.

A group of variables may be ordered into independent and dependent variables by ordering them in terms of time. Variables that come first in time are called independent variables. Variables that come later in time are called dependent variables. Figure 12.1 shows how all the variables in the detention home study may be ordered into a series of independent and dependent variables by using the time sequence principle.

Assuming that age and religion are determined at birth, they may be said to have occurred at the same time. Neither, then, may be called an independent or a dependent variable in relation to the other. However, both age and religion occur in time before any of the other variables. Age and religion may therefore be placed at the beginning of the time sequence. Note that they are placed parallel to one another at the far left of Figure 12.1. This indicates that they occurred at the same time.

Two variables now remain: seriousness of offense and days spent. Because placement in a detention home is usually the result of having committed an offense, seriousness of offense should precede in time the number of days spent in the detention home. Seriousness of offense is therefore placed after age and religion but before days spent.

Note that arrows have been drawn between some variables in Figure 12.1. The arrows between any two variables indicate which variable is hypothesized to be the independent variable and which is the dependent variable. The variable at the left of an arrow (at the arrow tail) is the independent variable. The variable at the right of an arrow (at the arrow head) is the dependent variable. When no arrow exists between two variables, as between age and religion, no predictive relationship is hypothesized.

Note as well that a variable may be an independent variable for several variables and that a variable may be both an independent and dependent variable. In the first case, age is an independent variable for both seriousness of offense and days spent. So is religion. In the latter case, seriousness of offense is a dependent variable in relation to age and religion but also an independent variable in relation to days spent. Overall, then, three variables are hypothesized to predict days spent (age, religion, and seriousness of offense), and two variables are hypothesized to predict seriousness of offense (age and religion) when the time sequence principle is used.

Use of the time sequence principle has thus helped to create a conceptual framework. The research problem is to explain the number of days spent in the detention home in terms of a client's personal characteristics (age and religion) and a client's behavior (seriousness of offense). From a theoretical perspective, this problem is important in understanding

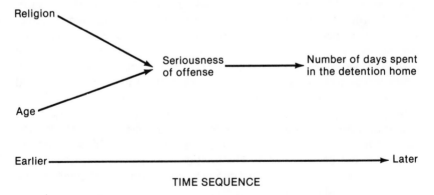

Figure 12.1. Ordering variables from the hypothetical detention home by the time sequence principle.

whether or not a client's personal characteristics and behavior influence the response of a youth services system. It might also suggest whether a youth services system such as a detention home reacts more to a client's personal characteristics or to the client's behavior in determining length of stay. From an applied perspective, this problem is important because, by knowing what predicts length of stay, it might be possible to design effective intervention strategies.

A literature review then summarizes previous findings on what variables predicted length of stay in detention homes or similar institutions, and suggests which hypotheses are most likely to be supported. Suppose, for example, that the literature review suggested the hypotheses shown in Figure 12.1:

1. Religion is hypothesized to be related to (predict) seriousness of offense.
2. Religion is hypothesized to be related to the number of days spent in the detention home.
3. Age is hypothesized to be related to seriousness of offense.
4. Age is hypothesized to be related to the number of days spent in the detention home.
5. Seriousness of offense is hypothesized to be related to the number of days spent in the detention home.

Further, because seriousness of offense is closest in time to days spent, another hypothesis might be added:

6. Of all the hypothesized relationships, the relationship between seriousness of offense and days spent is expected to be the largest.

After the hypotheses are stated, the methodology section is presented. The subjects were 30 girls in a detention home during January of some year. The research design consisted of a cross-sectional survey of all girls' files in the detention home during that month. Two variables were used for personal characteristics: a girl's age and religion. Religion was dichotomized to Protestant (a score of 1) and all others (a score of zero). One variable was used to describe a client's behavior: seriousness of offense. The seriousness of offense was ranked, let us suppose by a legal professional, on a three-point scale: 1 = low seriousness, 2 = moderate seriousness, and 3 = high seriousness. The number of days spent in the detention home was determined from the daily census reports for the home. The means and standard deviations for each variable are reported at the bottom of Table 12.1. The relationships between variables were measured by Pearson's product-moment correlation coefficient, r, and are shown in Table 12.1.

The empirical findings are then presented in a results section, perhaps something as follows for this particular study. The relationships

Table 12.1.   Pearson product-moment correlation matrix for all variables ($n = 30$)

	Age	Religion	Days spent	Offense
Age	1.00	−.23	.46	.62
Religion		1.00	−.14	−.04
Days spent			1.00	.40
Offense				1.00
Mean ($\overline{Y}$)	13.500	.50	7.167	1.900
Standard deviation (s)	1.925	.50	3.239	.845

between all variables are shown in the Pearson product-moment correlation matrix in Table 12.1. The correlation between religion and seriousness of offense is close to zero. Thus, hypothesis 1 is not supported. The correlation between religion and days spent is very low and negative. There is thus a slight trend for non-Protestant clients to spend more time in the detention home. Given the low level of association, however, hypothesis 2 cannot be said to be supported. The correlation between age and seriousness of offense is high and positive ($r = .62$). Hypothesis 3 is thus supported. As age increases, so does the seriousness of offense. The correlation between age and days spent is moderate and positive ($r = .46$). Hypothesis 4, therefore, is supported. As age increases, so does the number of days spent in the detention home. The correlation between seriousness of offense and days spent is moderate and positive ($r = .40$). Hypothesis 5 is thus supported. As the seriousness of offense increases, so does the number of days spent in the detention home. Lastly, because the correlation between age and days spent is slightly higher than the correlation between seriousness of offense and days spent, hypothesis 6 is not supported.

A discussion of these findings is the next step in a research report. From a theoretical perspective, it is clear that religion should be dropped from further consideration. Moreover, no more than 21 percent of the error was reduced in predicting days spent by an independent variable ($.46^2 = .21$ for age and days spent). Other variables should be added to the model to improve the prediction of length of stay. As such, the conceptual approach taken to explain length of stay in a detention home appears to require substantial modification.

What is known from this study is that age and seriousness of offense both predict the number of days spent in the detention home. What is not clear from this study, however, is why age is a predictor of days spent. Everyone grows older, but not everyone is placed in a detention home. Could it be that age is associated with seriousness of offense (there is strong empirical evidence to suggest this: $r = .62$)? Or is it that age is related to other variables not included in this study, such as a girl's previous record (an older girl is more likely to have a longer record and is thus more likely to commit serious offenses), judges' bias toward older girls (older girls should know better and thus should be punished more), or to the difficulty of placing older girls in other places like foster homes? The data analysis presented here does not answer these questions. Further research is required, therefore, to identify all relevant variables and to determine their predictive relationships. The need for further research is one reason why theory building takes so long.

From an applied perspective, the relationships between age, seriousness of offense, and number of days spent in the detention home suggest one plausible intervention strategy: intervene at the youngest age possible to prevent girls from committing more serious offenses and thus avoid the necessity for placing them in a detention home for a long period of time. A major difficulty with this intervention strategy is that it does not specify how this should be done, how likely the desired result actually is, or how long it will take to achieve the desired result. Moreover, the empirical data do not even allow for a comparison of the potential efficacy of several different strategies. In sum, therefore, although empirical research may suggest some type of intervention strategy, the practitioner should be aware that other strategies may be just as or more effective. Only further research allows an empirical basis for such decision making.

Such is a research report—a story that suggests what empirical information is important and a story whose ending is constrained by the empirical findings. The purpose of descriptive statistics is to provide a concise summary of the findings. Once learned, almost all the relevant material in Chapters 4 through 11 can be summarized in a table such as Table 12.1.

Part III of this book discusses ways in which the findings of this sample of girls in a detention home may be generalized to a larger population.

Chapter 13

Review

Two of the most important steps to review are the decision-making framework for descriptive statistics and the choice of appropriate statistics in a research report.

To retrace the decision-making framework, start with Figure 1 in the introduction to Part II. Study this decision-making chart for 2 minutes, close the book, and redo the chart by memory. Then write a one-page paper defining the intent or meaning of each box. Repeat this procedure for every decision-making chart in Chapters 5 through 11 (Figures 5.1, 6.1, 7.2, 8.2, 9.1, 10.1, and 11.1).

To interpret the choice of appropriate statistics in a research report, analyze the data in Table 13.1 and write a research report. This data list refers to 30 disabled adults who are either physically or mentally handicapped and who participate in a Center for Independent Living: a community-based, nonprofit, nonresidential program controlled by disabled consumers for the purpose of obtaining total integration of people with severe disabilities into community life (see Title VII, Comprehensive Service for Independent Living, of P.L. 95-602, The Rehabilitation, Comprehensive Services and Developmental Disabilities Amendments of 1978). The analysis should begin by formulating a conceptual framework as discussed in Chapter 12. The analysis should then proceed with univariate distributions for each variable along with their measures of central tendency and variability (Chapters 5 through 8). This analysis should then consider bivariate distributions for pairs of variables and their measures of relationships (Chapters 9 through 11). Finally, the research report should be written following the guidelines in Chapter 12. Before beginning, it might help to compare carefully the data list in Table 4.1 with the data list in Table 13.1.

After accomplishing each step, students should understand the logic of decision making in descriptive statistics and be able to communicate their knowledge through research reports. Both are skills highly respected and needed in the applied world—whether used to evaluate programs, write annual reports, develop funding proposals, supervise research projects, or participate in the research itself.

Table 13.1. A data list for handicapped persons participating in a center for independent living ($n = 30$)

Subject identification number	Number of years an advocate of rights for the handicapped	Type of handicap[a]	Number of months spent at the Center	Seriousness concerning independence[b]
1	11	1	5	1
2	17	3	11	3
3	14	1	8	1
4	13	3	5	1
5	13	2	10	3
6	14	2	9	3
7	14	2	12	2
8	16	2	13	3
9	13	1	7	1
10	12	1	4	1
11	15	3	3	2
12	10	1	8	2
13	15	3	9	2
14	14	1	5	2
15	11	2	1	1
16	12	2	6	1
17	14	1	9	2
18	15	1	2	3
19	16	1	14	3
20	13	2	9	1
21	15	2	4	3
22	14	1	11	3
23	17	1	8	3
24	16	2	7	2
25	12	1	4	2
26	13	1	7	1
27	13	2	9	1
28	12	2	6	1
29	11	1	5	1
30	10	1	4	2

[a]Type of handicap has three categories: 1 = physical, 2 = mental, 3 = both physical and mental.

[b]Seriousness concerning independence has three categories: 1 = low seriousness, 2 = moderate seriousness, and 3 = high seriousness.

13.1 SPSS

The *SPSS Primer* also contains data from two studies that may be analyzed for further review purposes. Read Appendix A of the *SPSS Primer* to become familiar with the data, then analyze the data, and lastly write research reports based on the data.

PART III

INFERENTIAL STATISTICS

Chapter 14

Basics for Hypothesis Testing

The purpose of data collection and data analysis is to describe and to infer. Descriptive analyses organize and summarize data in the sample. Inferential analyses guide the process of generalizing from the observed sample to a population. The difference between descriptive and inferential analyses rests on the way statistics are used and not on the statistics themselves. If statistics are used to summarize sample data for a discussion only of the sample, they are called descriptive statistics. If statistics based on sample information are used to estimate population values or parameters, they are called inferential statistics. Part II of this textbook focused on descriptive statistics. The focus of the remaining chapters is on inferential statistics.

Four basic concepts in inferential statistics will be discussed in this chapter: 1) simple random sampling, 2) sampling distributions, 3) estimation of population parameters, and 4) the logic of hypothesis testing. Specific hypothesis tests will be presented in the following chapter.

14.1 SIMPLE RANDOM SAMPLING

A simple random sample is one in which all subjects or elements (N) of the population are listed, and then n of these subjects are randomly selected to be part of the sample. The random selection is done in such a way that each subject has an *equal probability* or chance of being selected and that the choice of subjects is *independent*; that is, the choice of one subject does not influence the choice of any other subject. The easiest way to do this is: 1) write the list of all subjects in the population, 2) assign a number to each subject, for example, from 1 to the number of the last subject in the list, and 3) use the random numbers table in the Appendix (Table A) to select a simple random sample. To illustrate this procedure, let us suppose a sample of six subjects is to be drawn from a population of 1,000. Having compiled the list of 1,000 and having numbered them from 1 to 1,000, select any row and column number in Table A to start drawing the sample, for example, row 5 and column 1. In order to include numbers as large as 1,000 in the sample, columns 1 through 4 will need to be read.

Thus, the first number is 4,736, but because this number is larger than 1,000, it should be ignored. Proceeding downward from this starting point, the sample will include six subjects with the following numbers: 0875, 0370, 0297, 0523, 0784, and 0883. If the same number should be selected twice, general practice in the social sciences is not to include the same subject again in the sample. This procedure is called *sampling without replacement*. If this subject is included again in the sample, this procedure is called *sampling with replacement*. Strictly speaking, the definition of a simple random sample requires sampling with replacement. However, in practice, if the sample size (n) is small as compared to the population size (N), sampling without replacement makes very little difference.

One obvious reason for focusing on the concept of simple random sampling is that this type of sampling procedure most easily assures a representative sample of the population about which inferences are to be made. Any differences between sample and population characteristics, therefore, are due to random sampling errors or fluctuations and not to systematic differences. Another reason for focusing on simple random sampling is that this procedure is assumed for virtually all hypothesis tests. This assumption is made because it allows the sampling distributions of sample statistics to be easily defined.

14.2 SAMPLING DISTRIBUTIONS

A sampling distribution is a theoretical probability distribution of a random variable that: a) represents all possible values that might be obtained for a random variable, b) provides the probability or relative frequency with which these values would occur in the long run of drawing simple random samples of the same size from the same population, and c) the sum of probabilities is equal to 1.00; that is, the area under the graph of a sampling distribution is 100 percent.

A *random variable* is a variable for which a probability distribution is known and whose observed values are the outcome of random or chance factors. The word *variable* refers here to some sample statistic such as a mean, variance, or correlation rather than to a substantive variable such as age, race, or sex. The variable is called random because its exact value cannot be predicted in advance. Rather its particular value depends on the particular random sample drawn from a population, and there are a very large number of different random samples that may be selected. However, the relative frequency or probability with which each possible value of the random variable will occur in the long run is known at least theoretically in the context of a sampling distribution. The sampling distribution would have a mean and a standard deviation, called an *expected value* and a *standard error*, respectively, to distinguish them from a distribution for a substantive variable. For instance, if a very large number of simple random samples of the same size were drawn from the same population, and if sample means ($\overline{Y}$) were calculated for each of the samples, a sampling distribution of sample means would result. The average of all these sample means is called the *expected mean value* of the sampling distribution, and it is expected (and can be shown) to be equal to the true population mean. The standard deviation of all these sample means is called the *standard error of the mean* and is interpreted in the same way as a standard deviation (see Sections 8.6–8.8)—it indicates the probability of observing the specific sample statistic of the mean ($\overline{Y}$).

The sampling distribution of means may be described more generally by the *central limit theorem:* if random samples of size n are drawn repeatedly from any normally or non-normally distributed population with a population mean μ and a population variance σ^2, then, as the sample size increases, the sampling distribution of sample means approaches a normal distribution with a mean (expected mean value) of μ and a variance (standard error of means) of σ^2/n.

In sum, then, when some specific sample statistic is observed, the sampling distribution of that statistic indicates the likelihood of observing that value. Such sampling distributions—for means, variances, correlations, and so on—are the basis for hypothesis testing.

The hypothesis tests in this textbook make use of four theoretical sampling distributions: 1) the normal distribution, 2) the t distribution, 3) the chi-square distribution, and 4) the F distribution. Each sampling distribution defines the probability of observing a random variable of a certain size or larger. The general symbol for a random variable with a normal distribution is z, the general symbol for a random variable with a t distribution is t, the general symbol for a random variable with a chi-square distribution is χ^2, and the general symbol for a random variable with an F distribution is F. These random variables are transformations of such sample statistics as the mean, variance, and correlation coefficient observed in specific random samples. The transformation process results in what is called the test statistic in hypothesis tests. The likelihood of observing a specific value of a random variable (or a test statistic) is obtained from probability tables for each sampling distribution. These probability tables are in the Appendix: see Table C for the t and z random variables, Table D for the χ^2 random variable, and Table F for the F random variable. Each of these tables gives the value of a random variable that an observed test statistic must equal or surpass to be considered an extreme observation or an observation with a very low probability of occurring. This information is central to all hypothesis tests because it is the criterion for making statistical decisions.

Before discussing the logic of hypothesis tests, however, two last points about these sampling distributions are necessary. First, these sampling distributions are all closely related to each other in a mathematical sense. Although it is beyond the scope of this book to describe their relationship in any detail, one commonality is simply the assumptions required to derive these sampling distributions logically, namely, a) the assumption that subjects are selected from the population by simple random sampling, and b) the assumption that the population distribution of the substantive variable of interest forms a normal distribution. Second, the sampling distributions of t, χ^2, and F depend more on observed sample values than does the sampling distribution for z. That is, to use these distributions, more than one population parameter is estimated from sample data. The amount of error introduced by using additional estimates rather than known population parameters is accounted for in the use of degrees of freedom. The number of *degrees of freedom*, whose symbol is df, is equal to the number of observed scores minus the number of constraints placed on the scores. These constraints concern the number of observed rather than the previously known scores used to estimate a parameter. Specific directions on how to use and calculate degrees of freedom for each sampling distribution will be given in Chapter 15.

14.3   ESTIMATION OF POPULATION PARAMETERS

In inferential statistics, inferences are made about the population on the basis of sample information. The analysis of sample information results in values called statistics or observed values. These statistics are used to estimate population values called parameters or true values.

Assuming that simple random sampling has been used to assure that the sample is representative of the population, then two types of estimates may be used: point estimates and interval estimates. Both of these estimates are based on the concept of the sampling distribution of the population parameter of interest.

Point estimates use the best single sample value to estimate a parameter. Two of the most important ways to determine the "best" estimator in statistics are the criteria of:

1.   Unbiasedness—An estimate of a parameter is unbiased if the mean of its sampling distribution is equal to the population parameter.
2.   Efficiency—An estimate of a parameter becomes more efficient as the standard error of its sampling distribution becomes smaller.

An unbiased estimator denotes a point estimate that has the characteristic of being equal to the population value in the long run. Given a very large number of random samples of the same size, the average of the statistic calculated across samples will equal the population value. For example, the mean is an unbiased estimate of the true population mean, because the mean of the sampling distribution of a very large number of sample means calculated from random samples equals the true population mean. Thus, although a single sample mean may not equal the true population mean, over the long run the average of such sample means will. An example of a biased point estimate is the sample variance with a denominator of (n) rather than $(n-1)$. Only when the denominator is $(n-1)$ will the average of sample variances over the long run be equal to the true population variance; that is, the mean of the sampling distribution of variances will equal the true population variance. The denominator $(n-1)$ provides an unbiased estimate of the population variance. The reason is related to the concept of degrees of freedom. In order to calculate the sample variance, it is necessary to use a sample mean $(\overline{Y})$ about which the deviations of individual scores are calculated. Because the sample mean is an estimate of, rather than a known, population mean, one degree of freedom is lost; that is, one case was lost in computing the mean for the sample.

An efficient estimator denotes a point estimate that has a smaller standard error. A smaller standard error is preferred because it indicates less variability among sample estimates, and thus each sample estimate will be closer to the true population value. For instance, in a situation where the sample mean and median are equal, the standard error of the mean will be less than the standard error of the median. The mean, therefore, is the more efficient estimator.

Point estimates are the estimates used for hypotheses tests. However, a related concept is interval estimates. *Interval estimates* specify the degree of confidence you have that the specified interval around a point estimate actually includes the true value of a parameter. For this reason, interval estimates are called *confidence intervals*. Confidence intervals are seldom reported in social science literature in the human services, however. Consequently, they will not be discussed further except for the example given in Section 15.6.

14.4 HYPOTHESIS TESTING

Usually sample data are collected with the purpose of answering questions about a population. These questions are posed as hypotheses that present the research questions in terms of population parameters, because it is these population values that are of interest (the sample statistics are already known). The sample statistics are used, however, to estimate the population parameters. These sample statistics have known sampling distributions that are used to test hypotheses.

Although many hypotheses can be made, generally they focus on asking: Is this statistic (a distribution, mean, variance, or relationship) typical of a known population value, or is the difference between the same statistic observed in two or more groups large enough to indicate that these groups represent different populations? An example of the first question is the hypothesis that there is a correlation between height and weight. The sample correlation statistic is used as an estimate of the population correlation. This estimate is then compared against some known value in a sampling distribution to determine whether the difference between the estimated and the known value is due to random sampling fluctuations or to real population differences. An example of the second question is the hypothesis that the salaries of men and women differ. The difference between sample means of salaries for men and for women is used to estimate the population difference between means; this estimate is then compared to a value in a sampling distribution to determine whether the difference is due to random sampling fluctuations or to real population differences.

In statistics, a hypothesis test usually is done to evaluate a hypothesis you believe to be true. This is called the *research hypothesis*. For example, you believe that height is positively associated with weight, that women receive lower salaries than men for the same work, that authoritarian leaders are less likely to be concerned about the welfare of others than are altruistic leaders, that poor people have more health problems than wealthy people, and so on. Empirical support for the research hypothesis is achieved by rejecting a null hypothesis. The *null hypothesis* is defined to be the hypothesis about population values or parameters that will be rejected if the data are inconsistent with it but concurrently the data are consistent with the research hypothesis. The acceptance or rejection of the null hypothesis is based on probabilities determined by a statistic's sampling distribution. All hypothesis tests are based on the null hypothesis and all statistical decisions about the null hypothesis are based on a sampling distribution.

Statistical decisions about the null hypothesis use a level of significance to determine the region of rejection in a sampling distribution. A *level of significance*, whose symbol is alpha (α), is defined such that any observed statistic which falls into its range requires the null hypothesis to be rejected. The level of significance is usually set at .05 or .01. A .05 level of significance indicates that if an observed statistic is among the 95 percent of more likely outcomes, the null hypothesis is accepted. If, however, the observed statistic falls among the 5 percent of least likely outcomes, the null hypothesis is rejected and thus provides support for the research hypothesis. A .01 level of significance changes these percentages to 99 percent and 1 percent, respectively. The *region of rejection* is defined by the area of a sampling distribution that contains the 5 percent or 1 percent of these extreme values of the observed statistic. A *critical value* is the value that separates the region of acceptance from the region of rejection. A critical value of a test statistic is the value the statistic must equal or sur-pass to reach a given level of significance and thus to reject the null hypothesis.

A critical value is determined by: 1) a given level of significance, as already discussed; 2) the degrees of freedom, as already discussed; and 3) the type of research hypothesis. The *type of research* hypothesis usually determines whether a statistical test should be one-tailed or two-tailed. A research hypothesis may be of two types. It may only specify, for in-stance, that there is a correlation between two variables. Or the research hypothesis may further specify that the correlation is positive or negative. The former situation, in which the sign of the correlation is not specified, requires a two-tailed test; the latter situation, in which a direction is specified, requires a one-tailed test. A *two-tailed test* indicates that the region of rejection of a sampling distribution must consist of two parts: an area that contains extreme negative values and an area that contains extreme positive values. A *one-tailed test* indicates that the region of re-jection of a sampling distribution is only one area—either an area that contains extreme positive values or an area that contains extreme negative values. Figure 14.1 illustrates the difference between a one-tailed and a two-tailed test where the level of significance is .05. The top figure is of a sampling distribution and the regions of rejection for a two-tailed test. Here, .025 of the area in the left tail of the sampling distribution and .025 of the area in the upper or right tail represent the two regions of rejection that, when added together, are equal to the .05 level of significance. The lower two figures represent the region of rejection for one-tailed tests. The leftmost figure reflects a research hypothesis that specifies extreme positive values only for its empirical support, and thus .05 of the area in the right tail of the sampling is specified as the region of rejection. The rightmost figure reflects a research hypothesis that specifies only extreme negative values of the test statistic for its empirical support, and thus .05 of the area in the left tail of the sampling distribution is specified as the region of rejection. The advantage of one-tailed tests is that they have a greater likelihood of rejecting the null hypothesis and providing empirical support for the research hypothesis given the same value of a test statistic and the same level of significance. This happens because an observed test statistic may fall into an extreme .05 area of a sampling distribution, which for a one-tailed test would indicate that the null hypothesis should be rejected. However, if the test statistic had a probability of occurring between .05 and .025 (but not less than .025), then under a two-tailed test this would indicate the acceptance of the null hypothesis. Given this potential advantage of one-tailed tests, their use can only be justified if the researcher has a firm foundation for employing them. The two reasons that can justify this procedure are either previous knowledge

A Two-Tailed Test

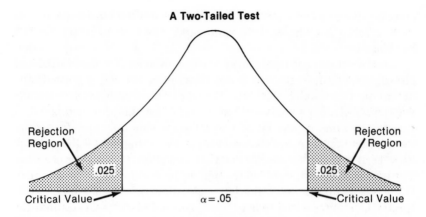

Right-Tail and Left-Tail Tests

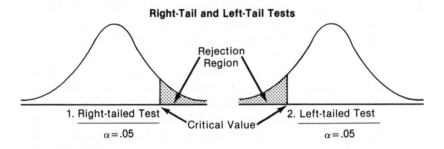

Figure 14.1.   Regions of rejection and critical values for hypothesis tests.

based on the results of past research or a logical deduction from a specified conceptual framework as to what should be expected.

Because hypothesis testing is based on sample information and not knowledge about the total population, the hypothesis-testing procedure cannot prove or disprove a null hypothesis. Rather, the procedure only provides information to indicate whether the sample result is sufficiently likely or unlikely to make the decision to accept or reject the null hypothesis. Thus, it is incorrect to say that the results "confirm" or "prove" a hypothesis. It can only be said that the results "provide support for" a hypothesis. The reason for this is based on an understanding of type I and type II errors.

Type I error is the rejection of a true null hypothesis. The probability of making a type I error is equal to the level of significance—which defines the region of rejection for a null hypothesis. Values of a test statistic that fall into the region of rejection of a sampling distribution are considered to be so extreme as to be very unlikely, and thus provide support for rejecting the null hypothesis. Yet given the definition of a sampling distribution, these extreme values could have been observed, if only rarely. Thus, if the level of significance is set at .05, 5 out of every 100 true null hypotheses will be rejected incorrectly. Type I error can be minimized by making the level of significance smaller; however, because type I and type II errors are inversely related, this procedure would increase the probability of a type II error. This is why a level of significance is usually set at .05 and less often at .01.

Type II error is the failure to reject a false null hypothesis; in other words, a decision is made to accept the null hypothesis when in fact the research hypothesis is true. It is very difficult to determine the probability of a type II error, but it can be minimized by increasing the sample size. The larger the sample becomes, the smaller is the risk of a type II error. If, however, the sample size is small (see Section 14.6), the level of significance should be set at .05 of higher, to lower the probability of a type II error.

14.5   THE HYPOTHESIS-TESTING PROCEDURE

There are generally seven steps in hypothesis testing. The *first* step is to state the null hypothesis. The null hypothesis is what is being subjected to the statistical test. The purpose of the test is to make inferences about population parameters; thus the null hypothesis is stated in terms of population parameters. The *second* step is to state the research hypothesis. The research hypothesis is based on a theory or conceptual framework. This statement should be as precise and complete as possible in order to make the later decision of using a one-tailed or a two-tailed test. This hypothesis is also stated in terms of population parameters. The remaining steps are all interrelated, but are differentiated here to define the hypothesis-testing procedure in detail. The *third* step is to choose a statistical test that fits the null hypothesis. Choice of the test should be made on the level of measurement of the variables and the assumptions necessary to use the test. The *fourth* step is identifying the sampling distribution of the test statistic under the null hypothesis. The *fifth* step is to know whether or not degrees of freedom are necessary to use this sampling distribution. The *sixth* step is to specify the assumptions underlying the use of the chosen sampling distribution. Although simple random sampling is an assumption common to the use of all sampling distributions, other assumptions may need to be made. These assumptions are listed for each hypothesis test in Chapter 15. The *seventh* step is to make a decision to accept or reject the null hypothesis. This decision involves the level of significance established to identify the region of rejection of a sampling distribution and the type of research hypothesis stated at the beginning of the procedure.

14.6 PARAMETRIC AND NONPARAMETRIC HYPOTHESIS TESTS

Hypothesis tests may be divided into two major groups: parametric tests and nonparametric tests. The use of parametric tests is based on two assumptions: 1) the variable or variables of research interest have a normal distribution in the population, and 2) the variable or variables are measured on an interval-level scale. The assumption of normality becomes less important as the size of the sample increases. For sufficiently large sample sizes, the sampling distributions of most statistics are essentially independent of the distribution of scores. For tests concerning means, a large sample size is about 30 or more (more as the departure from normality in the sample becomes greater); for tests concerning variances or Pearson r, a large sample size is about 50 to 100 or more. The *robustness* of parametric and nonparametric tests is defined as the accuracy of the probability levels assigned to test statistics on the basis of a sampling distribution. The robustness of parametric tests increases, then, with increases in sample size. The assumption about level of measurement is more constraining. Generally, when the level of measurement for a variable is nominal or ordinal, nonparametric tests are preferred. Although many nonparametric tests exist, only the chi-square test is defined in Chapter 15, because it is the most widely used nonparametric test.

The notion of increasing sample size to improve the robustness of a hypothesis test is related tangentially to another issue. Being able to reject a null hypothesis, and thus finding a statistically significant result, is a function of sample size as well as random sampling fluctuations. As the sample size increases, the critical value required to reject a null hypothesis decreases. Thus, it is quite possible to have trivial substantive findings but significant statistical results. Many solutions to this problem have been suggested; all solutions depend on common sense or good judgment criteria rather than statistical criteria. Common sense means looking at the descriptive statistics—their magnitude, size of differences between groups, and so on—to determine whether statistical significance represents a mountain or a molehill in the real, applied world. For example, it is quite possible to find that a Pearson r as low as .06 is statistically significant in a very large sample of about 750 people. Yet a Pearson r of .06 leaves over 99 percent of the variance unexplained. Thus, substantively speaking, little attention should be paid to this molehill even though it is statistically significant.

Chapter 15

Hypothesis Tests

This chapter presents 13 hypothesis tests. Although many more tests could have been included, these hypothesis tests represent tests that are used frequently in the human services. The focus of these tests is on univariate and bivariate analysis and thus closely parallels the discussion of descriptive statistics in Part II of this textbook. Problems of multivariate analysis—such as multiple regression, factor analysis, and multiple classification analysis of variance—are not discussed. However, the principles that underlie the tests presented here are the same as for multivariate analysis. Thus, once you are familiar with and understand these basic tests, the next step to multivariate analysis may be made.

Choice of which hypothesis test to use depends on 1) the purpose of the research, 2) the level of measurement for variables, 3) the number of groups or samples to be compared, 4) the type of comparison, and 5) the assumptions necessary to be able to use the sampling distribution of a test statistic. Figure 15.1 summarizes the path of choices given the first four concerns. The last concern about assumptions is discussed for each test separately.

1. The leftmost column of Figure 15.1 represents choices concerning the broad *purpose of the research:* to compare distributions, to compare means, to compare variances, and to compare relationships.

2. The intermediate columns represent choices on the basis of *measurement level of the variable(s).* Only two choices concerning level of measurement are included, whether the variable is nominal-level or interval-level.[1] A choice must be made in using this decision-making chart, then, whether to treat an ordinal-level variable as a nominal-level variable and thus lose information about rank ordering or to treat an ordinal-level variable somewhat inappropriately as an interval-level variable (see the rationale given in Chapters 7 and 11 for doing so). The choice concerning ordinal-level variables should also be considered in light of: 1) the specificity of the research question, that is, if the general question is about any unspecified difference, then nominal-level chi-square tests might be

---

[1]For excellent presentations of hypothesis tests for ordinal-level variables see Linton C. Freeman, *Elementary Applied Statistics* (New York: John Wiley & Sons, 1965) or Sidney Siegel, *Nonparametric Statistics* (New York: McGraw-Hill, 1956).

Figure 15.1.   Decision-making chart for hypothesis tests about populations using estimates from samples.

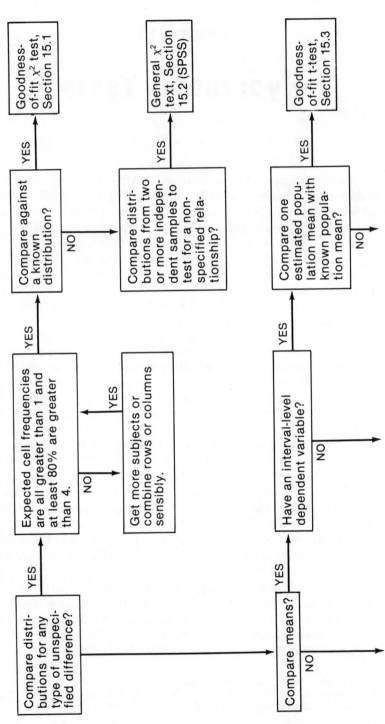

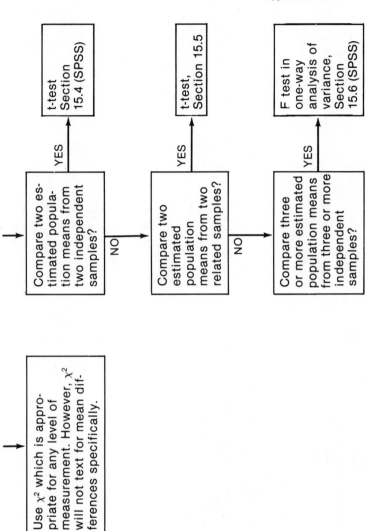

Figure 15.1 *continued*

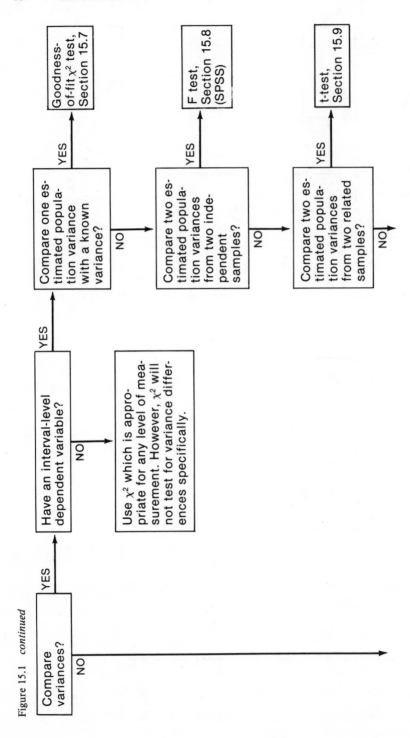

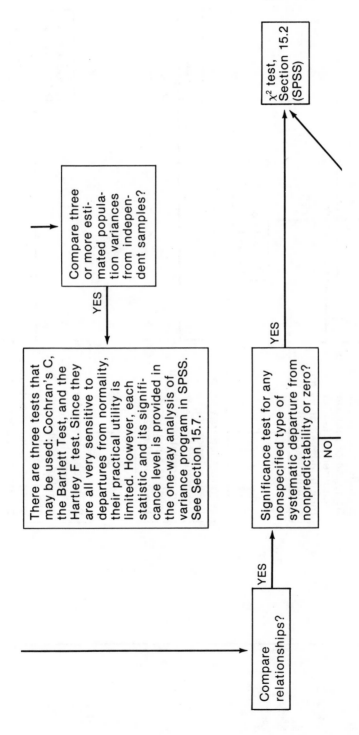

Compare three or more estimated population variances from independent samples?

YES

There are three tests that may be used: Cochran's C, the Bartlett Test, and the Hartley F test. Since they are all very sensitive to departures from normality, their practical utility is limited. However, each statistic and its significance level is provided in the one-way analysis of variance program in SPSS. See Section 15.7.

YES

Significance test for any nonspecified type of systematic departure from nonpredictability or zero?

YES

χ^2 test, Section 15.2 (SPSS)

NO

Compare relationships?

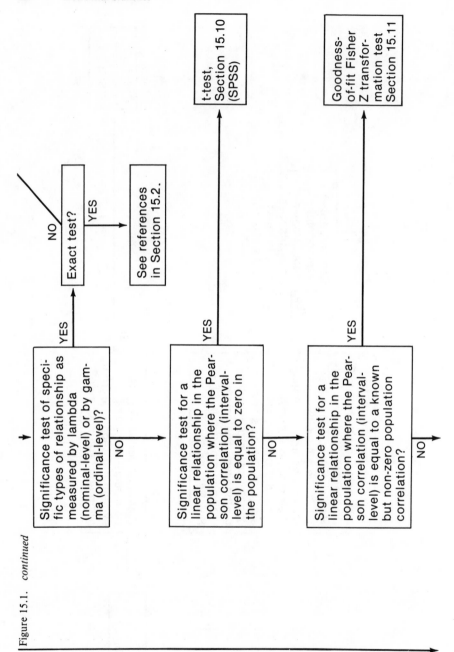

Figure 15.1. *continued*

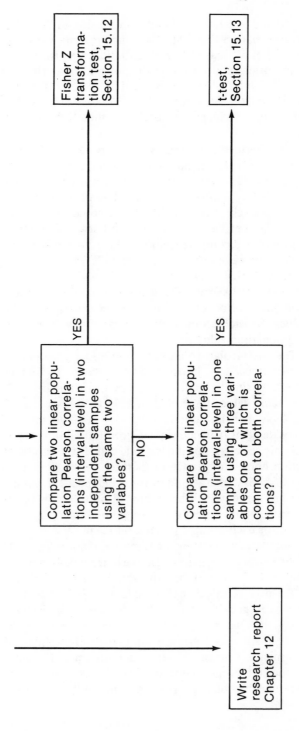

chosen, or if the question is about such specific differences as means, variances, and relationships, then the remaining interval-level tests might be chosen; and 2) the shape of the distribution, that is, if the shape is approximately normal and thus indicative of a normal distribution in the population, then interval-level tests might be chosen rather than nominal-level tests. In general, however, the conservative approach would be to use hypothesis tests designed for variables at their level of measurement or lower.

3.   The intermediate columns of Figure 15.1 also represent choices on the *number of groups* or samples to be compared. If only one group or sample is compared against an already known population, the comparison is called a *goodness-of-fit test*. The purpose of such tests is to determine how likely it is that the sample could have been drawn randomly from the specified population with known characteristics. If two or more groups are compared, all of which have unknown population characteristics or values estimated by sample values, the comparison is called a *test for two or more groups* or samples. The purpose of such tests is to determine how likely it is that all samples could have been drawn randomly from the same or common population; that is, the estimated population values for each group vary only randomly from a common but unknown population value.

4.   The intermediate columns of Figure 15.1 also indicate choices on the *type of comparison*, independent or related, to be made between groups. Because simple random sampling is assumed for all hypothesis tests, every score value of a variable may be treated as separate and *independent groups*. For the variable of sex, for example, males and females may be considered as two independent groups if all subjects in the sample were randomly selected. For the variable called seriousness of offense in the detention home study, each of its three values (low, moderate, and high) may be considered as one of three independent groups. These independent groups may then be compared on the basis of another variable, for example, days spent in the detention home, to see whether such estimates as for each group's population mean on days spent represent different population means or the same population mean. The variable used to define groups is usually called the independent variable (for example, seriousness of offense), and the variable used to describe potential group differences is called the dependent variable (for example, days spent). However, groups need not be independent but instead may be *related groups*. Related groups may come about for two reasons: 1) the same sample of subjects is measured at two different times, for example, staff performance is measured before and after a staff training program to see whether the program improved staff performance; or 2) pairs of subjects are matched on some characteristics and then tested for signifi-

cant differences on other variables, for example, a member of each pair is randomly assigned to an experimental medical treatment or to a control group receiving a placebo to see whether the medical treatment was effective. The purpose for using related rather than independent groups is to decrease the chance that true differences between groups will be obscured. Significant group differences are more likely to be observed for related groups rather than independent groups, because there should be more similarity of subjects in each group and thus fewer individual differences to introduce error.

5.   The rightmost column of Figure 15.1 indicates the name of the test chosen, the number of the section in this textbook that describes this test, and (in parentheses) whether SPSS has a program for this test.

Although the assumptions necessary to make the final decision are not all shown in Figure 15.1, they will be listed and discussed for each hypothesis test separately. These assumptions, which underlie the appropriate use of a test statistic's sampling distribution, may include one or more of the following: a) simple random sampling of subjects as discussed in Section 14.1; b) level of measurement as discussed above; c) number of groups to be compared and the type of group comparison as discussed above; d) normal distribution of scores on the dependent variable in the population or normal distribution of dependent variable scores within each group being compared in the population, as discussed in Sections 14.2 and 14.5; e) equality of population variances on the dependent variable in comparing each group or, in statistical terms, either the homogeneity of variances or the homoscedascity of variances; f) linear relationship between two variables in the population as discussed in Sections 11.4 and 11.7; g) bivariate normal distribution of joint scores on two variables in the population, such that the scores for each variable are normally distributed about scores for the other variable in the population; and h) a sufficiently large sample size, as assessed by expected frequencies in Section 15.1 and 15.2, to allow the distribution of a test statistic to be closely approximated by a specified sampling distribution.

Finally, the format for describing hypothesis tests is the same for each test. First, a *general description of the hypothesis* test is given. This description includes a discussion of the purpose of the test, a statement of the null hypothesis, a statement of the alternative research hypotheses, a presentation of the test statistic, the identification of the sampling distribution for this test statistic as well as its degrees of freedom, a statement of the assumptions needed to use this sampling distribution, and then a discussion of how to make the statistical decision. Second, after this general description, an *example* is given to demonstrate the use of each hypothesis test. Third, if there is an *SPSS program*, the procedure for doing the program and reading the printout is given. Lastly, several *com-*

ments may be added at the end of each hypothesis test section to discuss further the applied use or meaning of the test.

15.1 GOODNESS-OF-FIT CHI-SQUARE TEST FOR DISTRIBUTIONS

Purpose

The purpose of this test is to determine statistically whether an observed univariate distribution could have been selected randomly from a specified (previously known or theoretical) population distribution. That is, the frequencies for each cell of the observed distribution estimate the frequencies that would be expected if the sample was randomly drawn from the specified population. A *previously known distribution* means that the frequencies of some population have already been observed empirically. A *theoretical distribution* means that an assumption is made about what proportion of frequencies are expected in each cell. For example, an a priori assumption is made that an equal proportion of frequencies is expected in each cell.

Null Hypothesis

The observed frequencies (f_o) from a sample distribution could have been selected randomly from a population that has its distribution or expected frequencies (f_e) already specified.

$$H_0: \quad f_o = f_e.$$

Research Hypothesis

The observed frequencies from a sample distribution are not what would be expected on the basis of the specified population distribution. Thus, the sample distribution represents a different population from the population of the specified distribution (a two-tailed hypothesis).

$$H_R: \quad f_o \neq f_e.$$

Although one-tailed tests may be used if there are only two categories, the addition of more categories does not allow for a clear distinction about the direction of differences. Generally, then, only two-tailed tests are done.

Statistical Test of the Null Hypothesis

$$\chi^2 = \Sigma \frac{(f_o - f_e)^2}{f_e}$$

where

χ^2 = Pearson's chi-square test statistic, which is formulated as a random variable with a chi-square distribution
f_o = the observed frequency in each cell of the univariate distribution

f_e = the expected frequency for each cell of the univariate distribution. It is the frequency that is expected for each cell in the specified population distribution

$\dfrac{(f_o - f_e)^2}{f_e}$ = the discrepancy between the observed and expected frequencies (the greater the discrepancy, the more likely that the observed distribution represents a different population than in the specified distribution), which is squared and then divided by the expected frequency of that cell. The calculations for each cell are then summed across all cells (Σ) to provide the observed chi-square test statistic.

Sampling Distribution of This Test Statistic

Chi-square distribution.

Degrees of Freedom to Use This Sampling Distribution

$$df = k - 1$$

where

k is the number of columns in the univariate frequency distribution.

Assumptions Necessary to Use This Sampling Distribution

1. Simple random sampling is used to select subjects.
2. Variables are at least nominal-level measures and their categories are mutually exclusive and exhaustive.
3. The sample size must be sufficiently large, as assessed by expected frequencies; otherwise the chi-square sampling distribution does not provide accurate probabilities of occurrence for the test statistic.
4. The expected frequencies for cells are all greater than 1 and at least 80 percent are greater than 4. If this assumption is not met, then one of several choices must be made. First, increase the sample size to meet the assumption. Second, combine categories sensibly. For example, a) if there is a nominal-level variable concerning religion, combine all Protestant religions into one category; or b) if there is an ordinal-level variable with five categories, for example, strongly support, support, neutral, oppose, and strongly oppose, combine them into three categories, such as support, neutral, and oppose. Lastly, if one either begins with two categories or ends up with only two combined categories and still cannot meet the assumption, then the binominal test should be used.[2]

---

[2]See Siegel, *op. cit.*

Decision

Given a level of significance, a two-tailed research hypothesis, and the degrees of freedom, use the chi-square distribution in the Appendix (Table D) to make the statistical decision. If the observed chi-square statistic is less than the value in the table, accept the null hypothesis. If the observed chi-square statistic is equal to or greater than the value in the table, then there is support for the research hypothesis.

Example

Problem   A random sample of 80 children receiving public social services was selected. The reason for doing so was to compare current service utilization with that provided 10 years ago. One dimension of services of interest is the type of residence in which the child is living—in the family, in foster care or adoptive placement, in community group homes, or in institutions.

In the current sample of 100 children, the observed frequencies were: 25 children in homes, 40 in foster care or adoptive placement, 10 in group homes, and 25 in institutions.

In the survey done 10 years ago, the percentage of children observed in each category were: 25 percent in homes, 48 percent in foster care or adoptive placement, 5 percent in group homes, and 22 percent in institutions.

Question   Is the utilization pattern of residential services for children the same now as it was 10 years ago?

Hypotheses   The null hypothesis is that there is no difference in service utilization patterns or distributions. The research hypothesis is that service utilization patterns differ significantly; that is, the current sample represents a different population than what was found 10 years ago.

Procedure

1.   Arrange the data in a univariate frequency distribution.

Type of Residential Service

	Family care	Foster / Adoptive care	Group homes	Institutions
f_o	25	40	10	25
f_e	25	48	5	22

The observed frequencies are in the upper lefthand corner of each cell, and the calculated expected frequencies are in the bottom righthand corner. The observed frequencies are from the current sample of 100 children, and the expected frequencies are based on the survey of 10 years ago.

2. To calculate the expected frequencies for each cell, multiply the observed sample size by the percentage found in each cell in the previously known distribution. The expected frequency for the first cell is:

$$f_e = 100 \times 25\% = 25.$$

The expected frequency for the second cell is:

$$f_e = 100 \times 48\% = 48.$$

3. Calculate the test statistic:

$$\chi^2 = \frac{(25-25)^2}{25} + \frac{(40-48)^2}{48} + \frac{(10-5)^2}{5} + \frac{(25-22)^2}{22}$$

$$= 6.74.$$

4. Calculate the degrees of freedom: $df = 4 - 1 = 3$.
5. Given a level of significance of .05, a two-tailed research hypothesis, and 3 degrees of freedom, use Table D in the Appendix to make the statistical decision. Because the observed statistic of 6.74 is less than the table value of 7.815, accept the null hypothesis. This would be reported as $\chi_3^2 = 6.74$, $p > .05$.

SPSS

The *SPSS Primer* does not have this test, but it is available in the *Update 7-9* manual. For many situations, however, it is just as easy and fast to calculate the statistic by hand.

Comment

The above example uses a previously known empirical distribution as the specified comparison. A theoretical distribution may also be used. For example, suppose the assumption is made that each cell (or type of residential service) has an equal probability of being observed. This means that the percentage of frequencies is divided equally among cells. In the current example with four cells, each cell should have 25 percent of the frequencies and thus an expected frequency of 25 subjects in each cell (100 subjects $\times$ 25%). The resulting test statistic is $\chi^2 = 0 + 9 + 9 + 0 = 18.00$. Because the observed statistic of 18.00 is greater than the table value of 7.82 with 3 degrees of freedom, there is support for the research hypothesis. The observed distribution of services differs significantly from a distribution of services where it is assumed that all services are equally likely to be provided.

This goodness-of-fit test is a specific way to use the more general chi-square test for association presented below. The difference between them is that the goodness-of-fit test is for univariate or one-sample distributions and the general test is for bivariate or multisample distributions.

15.2   CHI-SQUARE TEST OF RELATIONSHIP

Purpose

This test determines statistically whether there is any unspecified type of relationship between two variables. It is most appropriate for nominal-level variables but may be used for ordinal- and interval-level variables as well. In Chapter 9, the analysis of bivariate contingency tables focused on the magnitude of differences between the groups defined by columns (scores on the independent variable) and across rows (scores on the dependent variable). The chi-square test evaluates whether these differences are large enough to support the conclusion that there are significant differences between these independent groups or samples; if so, this indicates a relationship between the two variables in the population.

Null Hypothesis

Two variables are unrelated to each other in the population, or the observed frequencies (f_o) do not differ by more than chance alone from the frequencies that would be expected (f_e) if there were no relationship in the population.

$$H_0: f_o = f_e.$$

Research Hypothesis

Two variables are related to each other in the population, or the difference between the observed and expected frequencies is so large that it suggests there is a relationship in the population.

$$H_R: f_o \neq f_e.$$

Generally, only a two-tailed research hypothesis is made, because the basic emphasis is on nominal-level measures where there is no distinction about direction.

Statistical Test of the Null Hypothesis

$$\chi^2 = \Sigma \frac{(f_o - f_e)^2}{f_e}$$

where

$\chi^2 =$ Pearson's chi-square test statistic which is formulated as a random variable with a chi-square distribution

$f_o =$ the observed frequency in each cell of the contingency table. It is the pattern of these observed frequencies that provides evidence of a relationship between variables.

$f_e =$ the expected frequency calculated for each cell in the contingency table. It is the frequency that is *expected* if there is no relationship between two variables in the population.

$\dfrac{(f_o - f_e)^2}{f_e}$ = the discrepancy between the observed and expected frequency of each cell of the contingency table (the greater the discrepancy, the more likely it is that there is a relationship between variables), which is squared and then divided by the expected frequency of that cell. The calculations for each cell are then summed across all cells (Σ) to provide the observed chi-square test statistic.

Sampling Distribution of This Test Statistic

Chi-square distribution.

Degrees of Freedom to Use This Sampling Distribution

$$df = (r - 1)(k - 1)$$

where

r = the number of rows that are in the contingency table
k = the number of columns that are in the contingency table.

Assumptions Necessary to Use This Sampling Distribution

1. Simple random sampling is used to select subjects.
2. Observations are independent; that is, observations on matched subjects cannot be used, nor can observations from the same subject at different times be used.
3. Variables are at least nominal-level measures.
4. The sample size must be sufficiently large, as assessed by expected frequencies; otherwise the chi-square sampling distribution does not provide accurate probabilities for the test statistic. When the degrees of freedom are greater than 1 ($df > 1$), at least 80 percent of the contingency table cells must have an expected frequency greater than 4, and no cell may have an expected frequency less than 1. If this assumption is violated, then rows and/or columns must be combined or more subjects need to be selected randomly from the population until this assumption is met. When the degrees of freedom are equal to 1 ($df = 1$, in other words for a 2 × 2 table), then one of two things must be done: 1) If the sample size is equal to or greater than 21, the chi-square test statistic requires Yates' correction for continuity in the numerator:

$$\chi^2 = \Sigma \frac{(|f_o - f_e| - .5)^2}{f_e}.$$

Before squaring the discrepancy for each cell, subtract .5 from the absolute (see Section 2.7) difference between f_o and f_e, then square the result. 2) If the sample size is less than 21, Fisher's exact test should be applied.[3]

---

[3]See Siegel, *op. cit.*

Decision

Given a level of significance, a two-tailed research hypothesis, and the degrees of freedom, use the chi-square sampling distribution in the Appendix (Table D) to make the statistical decision. If the observed chi-square statistic is less than the value in the table, accept the null hypothesis. If the observed chi-square statistic is equal to or greater than the value in the table, then there is support for the research hypothesis.

Example

Problem   A random sample of 400 children was selected from a public social services department. The children were classified into two groups: the handicapped (mentally disturbed, mentally retarded, and physically handicapped) and the nonhandicapped. Information was also collected on their current place of residence: in the family, in foster care or adoptive placement, or in group homes or institutions.

Question   Is there a relation between presence of handicap and the type of residence for handicapped chidren?

Hypotheses   The null hypothesis is that there is no statistically significant relationship in the population. The two-tailed research hypothesis is that there is some unspecified type of relationship between the presence of handicap and type of residence for these children in the population.

Procedure and Data

1. The data are arranged in a 2×3 contingency table where the observed frequencies are in the upper lefthand corner of each cell and the calculated expected frequencies are in the bottom righthand corner:

Handicap?

		Yes		No		
	Family	150		25		175
			131.2		43.8	
Type of	Foster/	120		40		160
residence	adoption		120.0		40.0	
	Group/	30		35		65
	institution		48.8		16.2	
		300		100		400

2. To calculate the expected frequencies for each cell, multiply the observed row total by the observed column total of that cell and then divide by the total sample size. The expected frequency for the cell in the first row and first column is:

$$f_e = \frac{175 \times 300}{400} = 131.2.$$

The expected frequency for the cell in the second row and first column is:

$$f_e = \frac{160 \times 300}{400} = 120.0.$$

3. Calculate the test statistic:

$$\chi^2 = \frac{(150 - 131.2)^2}{131.2} + \frac{(120 - 1200)^2}{120.0} + \frac{(30 - 48.8)^2}{48.8} +$$

$$\frac{(25 - 43.8)^2}{43.8} + \frac{(40 - 40)^2}{40} + \frac{(35 - 16.2)^2}{16.2} + \ = 49.82.$$

4. Calculate the degrees of freedom: $df = (3 - 1)(2 - 1) = 2$.
5. Given the level of significance at .05, a two-tailed research hypothesis, and 2 degrees of freedom, use Table D in the Appendix to make the statistical decision. Because the observed statistic of 39.82 is greater than the table value of 5.991, this provides support for the research hypothesis. This would be reported as $\chi^2 = 39.82$, $p < .05$. There is a statistically significant relationship between presence of a handicap and type of current residence. The asymmetric lambda, with the presence of a handicap as the independent variable and place of residence as the dependent variable, is .07. That is, only 7 percent of the errors of prediction concerning type of residence are accounted for by taking the presence of handicap into account. Thus, although there is a significant relationship between these variables (χ^2), the strength of the relationship is very low (lambda) and consequently not meaningful in a pragmatic sense.

SPSS

The CROSSTABS program calculates this test statistic and its significance level or probability of occurring using the chi-square sampling distribution (review Section 9.5 and Table 9.6). This program will automatically use Yates' correction for continuity or Fisher's exact test for 2×2 contingency tables. Table 9.6 shows that, without the correction for continuity, the raw chi-square is overestimated and thus provides an inaccurate (or too low an) estimate of the probability level. Without the correction for continuity, the observed test statistic provides false statistical support for the research hypothesis; with the correction, the observed statistic supports the null hypothesis. The researcher will make the correct decision only if the correct procedure is followed.

Comment

There are exact tests for testing the statistical significance of lambda (a measure of relationship for two nominal-level variables) and gamma (a measure of relationship for two ordinal-level variables). However, SPSS does not provide these specific hypothesis tests. Instead, the chi-square statistic is used to indicate that *any* type (not just the types specified by lambda or gamma) of systematic predictability or association exists in a population. If there is a complete lack of association of any type as indicated by chi-square, lambda and gamma will equal zero; if there is a complete association between variables, lambda and gamma will be 1.00. It is quite possible, however, that the chi-square statistic will indicate that some association exists, even though lambda or gamma is zero. As such, chi-square is not an exact test of either measure of relationship. It is common practice, however, to use chi-square as a significance test for lambda and sometimes (as in SPSS) for gamma. If this nonexact approach is taken, then the null hypothesis for lambda would be $\lambda = 0$ and for gamma would be $\gamma = 0$; the research hypothesis for lambda would be $\lambda > 0$ (lambda does not indicate direction because it is used for nominal-level variables) and that for gamma would be $\gamma \neq 0$.[4]

15.3   GOODNESS-OF-FIT *t*-TEST FOR A MEAN

Purpose

This test determines statistically whether a population mean (μ_1) estimated from an observed sample mean (Y_1) is equal to a specified population mean (μ). That is, the test asks whether the estimated population mean could have been obtained in a random sample from the same population with the specified population mean. This is, therefore, a goodness-of-fit test. A specified population mean is a mean that is either known from prior research or that is based on a theoretical assumption about what mean should be expected.

Null Hypothesis

The population mean (μ_1) estimated from a sample mean is equal to a specified population mean (μ).

$$H_0: \mu_1 = \mu.$$

Research Hypothesis

1. The population mean estimated from a sample mean differs from, or is not equal to, a specified population mean (a two-tailed test)

$$H_R: \mu_1 \neq \mu.$$
or

---

[4]An exact test for lambda is given in H. T. Reynolds, *Analysis of Nominal Data* (Beverly Hills, Calif.: Sage, 1977). An exact test for gamma is given in Freeman, *op. cit.*

2. The population mean estimated from a sample mean is greater than the specified population mean (a one-tailed test)

$$H_R: \mu_1 > \mu$$

or

3. The population mean estimated from a sample mean is less than the specified population mean (a one-tailed test)

$$H_R: \mu_1 < \mu.$$

Statistical Test of the Null Hypothesis

$$t = \frac{\overline{Y}_1 - \mu}{s_{\overline{y}}} = \frac{\overline{Y}_1 - \mu}{s/\sqrt{n}}$$

where

$t =$ a t-test which as formulated is a random variable with a t distribution. When there is little or no difference between $\overline{Y}_1$ and μ (that is, random sampling fluctuations could account for the difference), t will be equal or close to zero. This is what is expected under the null hypothesis. As the difference between $\overline{Y}$ and μ increases, so will the size of t increase given a constant standard error ($s_{\overline{y}}$).

$\overline{Y}_1 =$ the sample mean which estimates a population mean (μ_1)

$\mu =$ the specified population mean

$s =$ the sample standard deviation which estimates the standard deviation of the specified population, where $s = (\Sigma(Y_i - \overline{Y})^2)/(n-1)$

$n =$ the size of the sample

$s/\sqrt{n} =$ the estimated standard error of the mean whose symbol is $s_{\overline{y}}$. If a very large number of random samples of the same size (n) were drawn from the same population and a sample mean was calculated for each sample, the result would be a normal distribution of means. This normal sampling distribution of means has a mean (μ, which under the null hypothesis is equal to μ_1) and a standard deviation of means called the standard error ($\sigma_{\overline{y}} = \sigma/\sqrt{n}$). The standard error indicates the number of standard deviations in a sampling distribution between the population mean μ and the estimated population mean μ_1. Because σ is not known, however, it must be estimated from a sample standard deviation s. It follows, then, that the unknown population standard error $\sigma/\sqrt{n}$ must also be estimated by the sample standard error $s/\sqrt{n}$. This shift from a known to an unknown population standard deviation, which represents a loss of information and an introduction of more random sampling errors or fluctuations, means that the t sampling distribution rather than the normal sampling distribution should be used to evaluate the t-test statistic. This is discussed more in a comment at the end of this section.

Sampling Distribution of This Test Statistic

t distribution.

Degrees of Freedom to Use This Sampling Distribution

$n-1$.

Assumptions Necessary to Use This Sampling Distribution

1. Simple random sampling is used to select subjects.
2. The variable used for the calculation of the mean is an interval-level measure.
3. The population of the variable is normally distributed.

Decision

Given a level of significance, the type of research hypothesis, and the degrees of freedom, use Table C in the Appendix to make the statistical decision. For the *two-tailed* research hypothesis (hypothesis 1), if the absolute value (see Section 2.7) of the *t* statistic is equal to or greater than the table value, then there is support for the research hypothesis. Otherwise, accept the null hypothesis. For the *one-tailed* research hypothesis 2, if $\overline{Y}_1$ is greater than μ and if the observed *t* statistic is equal to or greater than the table value, then there is support for the research hypothesis. Otherwise, accept the null hypothesis. For the *one-tailed* research hypothesis 3, if $\overline{Y}_1$ is less than μ and if the absolute value of the observed *t* statistic is equal to or greater than the table value, then there is support for the research hypothesis. Otherwise, accept the null hypothesis.

Example

Problem   The executive director of an agency that provides group counseling set the goal of having an average of nine people ($\mu = 9$) in each group session. After the research question was asked, a random sample of 16 group sessions showed that there was an average of 10 people ($\overline{Y}_1 = 10$) in each group session and that the sample standard deviation was 2.0 people ($s_1 = 2.0$).

Question   Does agency practice, in terms of size of group sessions, differ significantly from the expectations of the executive director?

Hypotheses   The null hypothesis is that there is no difference between the executive director's expectations and agency practice in terms of the size of group sessions. The two-tailed research hypothesis is that there is a significant difference between expectations and practice.

Procedure

1. Calculate the *t* statistic:

$$t = \frac{10-9}{2.0/\sqrt{16}} = 2.000.$$

2. Calculate the degrees of freedom: $df = 16 - 1 = 15$.
3. Given a level of significance at .05, a two-tailed research hypothesis, and 15 degrees of freedom, use Table C in the Appendix to make the

statistical decision. Because the observed t statistic of 2.000 is less than the table value of 2.131, accept the null hypothesis. This would be reported as $t_{15} = 2.131, p > .05$.

SPSS

Because SPSS does not provide a program for this test, it must be calculated by hand.

Comment

In discussing the above test, it was assumed that the specified population mean (μ) was known but that the population variance (σ) was not. However, if the population variance were also known, then a z test rather than a t test would be more appropriate. When more is known, the less likely it is that significant differences will be obscured. The additional knowledge provided by knowing σ allows for more precise decision making, simply because σ is known and therefore need not be estimated by s. In the above example, the null hypothesis was accepted. If it were known, however, that the population standard deviation was 2.0 ($\sigma = 2.0$) rather than only estimated to be 2.0 ($s = 2.0$), then the null hypothesis would have been rejected and there would have been support for the research hypothesis, because the critical value of z was ± 1.96.

The null hypothesis, the options for alternative research hypotheses, and the assumptions do not change in shifting from the t test to the z test, but the test statistic and its sampling distribution do change. The z test statistic is:

$$z = \frac{\overline{Y}_1 - \mu}{\sigma / \sqrt{n}}.$$

Note that the change here is to use σ rather than s. The sampling distribution of z is the standard normal distribution (no degrees of freedom are necessary). Decision making is based on the z values given in Table C of the Appendix—given a level of significance and the type of research hypothesis—and follows the same logic as for the t test. For a *two-tailed* research hypothesis, as in the current example, since the observed z of 2.000 is greater than the table value of 1.960, there is support for the research hypothesis. Thus, being able to specify more knowledge—both μ and σ—allows the researcher to detect significant differences that may be obscured when using estimates and therefore less precise knowledge. When n becomes large (again more knowledge is added by sampling more subjects), however, the distribution of t will approach the standard normal distribution, and the probabilities associated with a t statistic can be approximated by the normal probabilities. The advantage of the z test, therefore, is less apparent as the sample size becomes large.

15.4   *t*-TEST FOR COMPARING POPULATION MEANS FROM TWO INDEPENDENT SAMPLES OR GROUPS

Purpose

This test determines statistically whether two population means (μ_1 and μ_2) estimated from sample information ($\overline{Y}_1$ and $\overline{Y}_2$) are equal. Both means are from independent samples or groups.

Null Hypothesis

The population mean of group 1 (μ_1) is equal to the population mean of group 2 (μ_2).

$$H_0: \mu_1 = \mu_2.$$

Research Hypothesis

1. The population mean of group 1 differs from or is not equal to the population mean for group 2 (a two-tailed test)

$$H_R: \mu_1 \neq \mu_2$$
or

2. The population mean of group 1 is greater than the population mean of group 2 (a one-tailed test)

$$H_R: \mu_1 > \mu_2$$
or

3. The population mean of group 1 is less than the population mean of group 2 (a one-tailed test)

$$H_R: \mu_1 < \mu_2.$$

Statistical Test of the Null Hypothesis

$$t = \frac{(\overline{Y}_1 - \overline{Y}_2)}{s_{(\overline{Y}_1 - \overline{Y}_2)}} = \frac{(\overline{Y}_1 - \overline{Y}_2)}{\sqrt{\dfrac{(n_1 - 1)s_1^2 + (n_2 - 1)s_2^2}{(n_1 - 1) + (n_2 - 1)}} \sqrt{\dfrac{1}{n_1} + \dfrac{1}{n_2}}}$$

where

t = a *t*-test which as formulated is a random variable with a *t* distribution. When there is little or no difference between the estimated population means $\overline{Y}_1$ and $\overline{Y}_2$ (that is, random sampling fluctuation could account for the difference), *t* will be equal or close to zero. As the difference increases, so will the size of *t* increase given a constant standard error ($s_{\overline{Y}_1 - \overline{Y}_2}$).

$\overline{Y}_1$ = a sample mean which estimates the population mean for group 1 (μ_1)

$\overline{Y}_2$ = a sample mean which estimates the population mean for group 2 (μ_2)

s_1^2 = a sample variance which estimates the population variance for group 1 (σ_1^2)

s_2^2 = a sample variance which estimates the population variance for group 2 (σ_2^2)

n_1 = sample size for group 1

n_2 = sample size for group 2

$S_{\bar{Y}_1 - \bar{Y}_2}$ = the estimated standard error of mean differences. The differences of sample means have been found to be normally distributed, with a population mean of $(\mu_1 - \mu_2)$ and a population variance of $(\sigma_1^2/n_1 + \sigma_2^2/n_2)$. Under the null hypothesis, this population mean is assumed to be zero $(\mu_1 - \mu_2 = 0)$. The square root of this population variance is called the standard error or the standard deviation of mean differences:

$$\sigma_{\bar{y}_1 - \bar{y}_2} = \sqrt{\frac{\sigma_1^2}{n_1} + \frac{\sigma_2^2}{n_2}}.$$

If it is assumed (see assumption 4 below) that σ_1^2 equals σ_2^2, or in other words that both equal σ, then:

$$\sigma_{\bar{y}_1 - \bar{y}_2} = \sigma\sqrt{\frac{\sigma^2}{n_1} + \frac{\sigma^2}{n_2}} = \sigma\sqrt{\frac{1}{n_1} + \frac{1}{n_2}}.$$

The common population standard deviation σ can now be estimated by using a "pooled" estimate (S_{pooled}) based on both sample standard deviations. This pooled estimate is under the first radical sign of the denominator of the t-test and is weighted by sample size (n_1 and n_2) to take account of different sample sizes for group 1 and group 2:

$$S_{pooled} = \sqrt{\frac{(n_1 - 1)s_1^2 + (n_2 - 1)s_2^2}{(n_1 - 1) + (n_2 - 1)}}.$$

Sampling Distribution of This Test Statistic

t distribution.

Degrees of Freedom to Use This Sampling Distribution

$$df_{pooled} = (n_1 - 1) + (n_2 - 1) = n_1 + n_2 - 2.$$

Assumptions Necessary to Use This Distribution

1. Simple random sampling is used to select subjects.
2. The variable used for the calculation of the mean is an interval level measure.
3. The distribution of the variable is normal in both populations or groups.
4. The population variances of the variable are equal in both populations $(\sigma_1^2 = \sigma_2^2)$; that is, in statistical language, homoscedasticity or homogeneity of variances is assumed.

Decision

Given a level of significance, the type of research hypothesis and the degrees of freedom, use Table C in the Appendix to make the statistical decision. For the *two-tailed* research hypothesis (hypothesis 1), if the absolute value (see Section 2.7) of the t statistic is equal to or greater than the table value, then there is support for the research hypothesis. Otherwise, accept the null hypothesis. For the one-tailed research hypothesis 2, if $\overline{Y}_1$ is greater than $\overline{Y}_2$, and if the observed t statistic is equal to or greater than the table value, then there is support for the research hypothesis. Otherwise, accept the null hypothesis. For the *one-tailed* research hypothesis 3, if $\overline{Y}_1$ is less than $\overline{Y}_2$, and if the absolute value of the observed t statistic is equal to or greater than the table value, then there is support for the research hypothesis. Otherwise, accept the null hypothesis.

Example

Problem   Sheltered workshops provide job training for disabled persons. The quality of the training depends in part on the ratio of trainees to the number of work trainers or supervisors. It is presumed that the fewer the trainees per supervisor, the better the quality of training. A national sample of sheltered workshops was randomly selected to see whether better trainee/supervisor ratios existed in unionized (group 1) or nonunionized (group 2) workshops. The data provided the following information:

Group 1	Group 2
$n_1 = 20$	$n_2 = 42$
$\overline{y}_1 = 8$ clients to one supervisor	$\overline{y}_2 = 10$ clients to one supervisor
$s_1^2 = 9$	$s_2^2 = 10$

Question   Do unionized and nonunionized sheltered workshops differ in trainee/supervisor ratios?

Hypotheses   The null hypothesis is that there is no mean difference between unionized and nonunionized workshops in the population concerning the variable of trainee/supervisor ratios. The two-tailed research hypothesis is that there is a mean difference between unionized and nonunionized workshops in the population concerning this variable.

Procedure

1. Calculate the test statistic:

$$t = \frac{(8-10)}{\sqrt{\dfrac{(20-1)9 + (42-1)10}{(20-1) + (42-1)}} \sqrt{\dfrac{1}{20} + \dfrac{1}{42}}}$$
$$= -2.380.$$

2. Calculate the degrees of freedom: $df = (20-1) + (42-1) = 60$.
3. Given a level of significance of .05, a two-tailed research hypothesis and 60 degrees of freedom, use Table C in the Appendix to make the statistical decision. The absolute value of 2.380 for the t statistic is greater than the table value of 2.000; thus there is support for the

research hypothesis. This would be reported as $t_{60} = -2.38$, $p < .05$. Because the union mean is lower than the nonunion mean, this test indicates that this aspect of quality of training (the trainee/supervisor ratio) is better in unionized rather than in nonunionized sheltered workshops.

SPSS

The SPSS program for this t-test is called T-TEST. See Chapter 11 in the *SPSS Primer*. An example from the detention home study, described in Part I, will demonstrate how to use this program.

The 30 girls in the detention home study are divided into two groups on the basis of the "seriousness of offense" variable. Group 1 consists of 12 girls who have a low score (1) on this variable; group 2 consists of 18 girls who have an intermediate (2) or high score (3) on this variable. This recoding of seriousness of offense is done by the following SPSS statement:

```
1                  16
RECODE             OFFENSE (1 = 1) (2,3 = 2)
```

The question is whether girls in group 1 differ from girls in group 2 in terms of the dependent variable called days spent in the detention home. This question indicates that a two-tailed research hypothesis would be appropriate where $\mu_1 \neq \mu_2$. The null hypothesis ($\mu_1 = \mu_2$) may be tested by the t test in this section.

This t-test is accomplished in SPSS by the following card:

```
1                  16
T-TEST             GROUPS = OFFENSE (1,2)/VARIABLES = DAYS
```

The "GROUPS = " specification indicates what variable is used to divide the sample into two groups. Here the variable is OFFENSE, which has two groups labeled 1 and 2. The "VARIABLES = " specification indicates what variable is to be the dependent variable or the variable on which mean differences are predicted.

Table 15.1 demonstrates what the SPSS printout would look like. At the upper left, the independent or grouping variable is printed, for example, GROUP 1 – OFFENSE EQ 1. Under the VARIABLE column, the name of the dependent variable (DAYS) is printed. The table is then divided into two rows: one row for GROUP 1 and the other row for GROUP 2. The NUMBER OF CASES column then indicates how many subjects are in each group: 12 in group 1 and 18 in group 2. The MEAN column indicates the sample mean for each column: 6.0 days for group 1 and 7.9444 days for group 2. The STANDARD DEVIATION column indicates the estimated population standard deviation for each group: 2.256 days for group 1 and 3.605 days for group 2. The next columns of immediate interest are those under the heading POOLED VARIANCE

Table 15.1 A printout for a *t*-test comparing means from two independent groups

GROUP 1—OFFENSE EQ   1.
GROUP 2—OFFENSE EQ   2.

							POOLED VARIANCE ESTIMATE			SEPARATE VARIANCE ESTIMATE		
VARIABLE	NUMBER OF CASES	MEAN	STANDARD DEVIATION	STANDARD ERROR	F VALUE	2-TAIL PROB.	T VALUE	DEGREES OF FREEDOM	2-TAIL PROB.	T VALUE	DEGREES OF FREEDOM	2-TAIL PROB.
DAYS NUMBER OF DAYS IN DETENTION												
GROUP 1	12	6.0000	2.256	0.651	2.55	2.55	−1.66	28	.108	−1.82	27.94	0.080
GROUP 2	18	7.9444	3.605	0.850								

ESTIMATE. The T-VALUE column reports the t statistic of -1.66, which is calculated as follows:

$$t = \frac{6.0 - 7.9444}{\sqrt{\dfrac{(12-1)(2.256^2)+(18-1)(3.605^2)}{(12-1)+(18-1)}} \sqrt{\dfrac{1}{12}+\dfrac{1}{18}}}.$$

The DEGREES OF FREEDOM column reports the pooled degrees of freedom as 28, calculated as $(12-1)+(18-1)$. The 2-TAIL PROB. column reports the probability of observing a t statistic of this size, with 28 degrees of freedom and assuming a two-tailed research hypothesis. If the level of significance were set at .05, then a .108 probability of observing this t statistic leads to an acceptance of the null hypothesis because this probability is greater than .05. One can doublecheck this by using Table C in the Appendix—the observed t statistic of -1.66 with 28 degrees of freedom is less than the absolute value for a two-tailed test (2.048). If the research hypothesis were one-tailed, then the two-tailed probability should be divided in half (.108/2) to provide the one-tailed probability.

One assumption underlying the t-test is the equality of population variances between groups. The F test in Section 15.8 is used to test this assumption. This is what the F VALUE column and its adjacent column of 2-TAIL PROB. are reporting. Because these results indicate that the estimated population variances for the two groups are equal, the pooled variance estimate (s_{pooled}) may be used to calculate the t-test. However, if the F test had indicated that the population variances were not equal, then the set of columns under the heading of SEPARATE VARIANCE ESTIMATE should be used to determine the significance of the t-test. The separate variance estimate refers to the STANDARD ERROR of the mean ($s_1/\sqrt{n_1}$ and $s_2/\sqrt{n_2}$) for each group (.651 and .850, respectively). It is beyond the scope of this text to discuss this alternative process, but, generally speaking, the decision-making process is the same as described for the pooled variance estimate procedure.

15.5 t-TEST FOR COMPARING TWO
POPULATION MEANS FROM RELATED SAMPLES OR GROUPS

Purpose

This test determines statistically whether the mean difference (μ_D) between two population means (μ_1 and μ_2) is zero when these means are estimated ($\overline{Y}_1$ and $\overline{Y}_2$) from related samples or groups. *Related samples* may be of two kinds: a) the same subjects are measured before and after an intervention, or b) pairs of subjects have been matched on relevant characteristics, and one member of each pair is randomly assigned to an experimental group and the other member of the pair is assigned to the control group.

The purpose of this test is essentially the same as described for the t-test in Section 15.4, except that the use of related samples may reduce the sampling error (standard error of mean differences) and thus increase the likelihood that a real mean difference between groups will not be obscured. The sampling error may be reduced because using subjects as their own controls (method a) or randomly assigning matched pairs (method b) may remove sources of variability within pairs (that is, differences due to subject characteristics) that might affect the estimated group means.

Null Hypothesis

The difference between population means from related samples is zero.

$$H_0: \mu_D = \mu_1 - \mu_2 = 0.$$

Research Hypothesis

1. The difference between population means from related samples differs from or is not equal to zero (a two-tailed test)

$$H_R: \mu_D \neq 0$$
or

2. The difference between population means from related samples is greater than zero (a one-tailed test)

$$H_R: \mu_D > 0$$
or

3. The difference between population means from related samples is less than zero (a one-tailed test)

$$H_R: \mu_D < 0.$$

Statistical Test of the Null Hypothesis

$$t = \frac{\overline{Y_d} - \mu_D}{s_d}$$

where

t = a t-test which as formulated is a random variable with a t distribution. When there is little or no difference between the means from related groups ($\overline{Y}_1$ and $\overline{Y}_2$), t will be equal or close to zero; that is, random sampling fluctuations could account for the differences. As the mean difference increases, so will the size of t increase with everything else held constant.

$\overline{Y}_d$ = the difference between the sample means of the related groups ($\overline{Y}_1 - \overline{Y}_2$)

μ_D = the difference between the population means of the related groups ($\mu_1 - \mu_2$). This difference, under the null hypothesis, is assumed to be zero.

s_d = the estimated standard error for population mean differences between related samples. It is calculated as follows:

$$s_d = \sqrt{\frac{s_1^2 + s_2^2 - 2r_{12}s_1s_2}{n}}.$$

s_1^2 and s_2^2 are estimated population variances for each group; s_1 and s_2 are estimated population standard deviations for each group; n is the number of pairs of subjects (not the total number of subjects here); and r_{12} is the Pearson correlation between the dependent variable for pairs of subjects. The term $r_{12}s_1s_2$ is the covariance between the variables Y_1 and Y_2. The larger the covariance, with everything else held constant, the more likely it is that a significant difference between means will be found.

Sampling Distribution of This Test Statistic

t distribution.

Degrees of Freedom to Use This Sampling Distribution

$n - 1$, where n is equal to the number of pairs.

Assumptions Necessary to Use This Distribution

1. Simple random sampling is used to select subjects for a before-and-after test, or members of previously matched pairs are randomly assigned to the experimental and control groups.
2. The variable used for the calculation of the means is an interval-level measure.
3. The distribution of the variable in each population is normal, and thus the mean differences are normally distributed in the population.

Decision

Given a level of significance, the type of research hypothesis and the degrees of freedom, use Table C in the Appendix to make the statistical decision. For a two-tailed research hypothesis (hypothesis 1), if the absolute value (see Section 2.7) of the t statistic is equal to or greater than the table value, then there is support for the research hypothesis. Otherwise, accept the null hypothesis. For the one-tailed research hypothesis 2, if the sample mean difference is greater than zero, and if the observed t statistic is equal to or greater than the table value, then there is support for the research hypothesis. Otherwise, accept the null hypothesis. For the one-tailed research hypothesis 3, if the sample mean difference is less than zero, and if the observed t statistic is equal to or greater than the table value, then there is support for the research hypothesis. Otherwise, accept the null hypothesis.

Example

Problem A job training program for women on public welfare is being initiated. The director of the program wants to determine whether the program results in reducing the amount of welfare payments to these women. It is expected that the program will enable these women to get jobs and a salary, which would consequently reduce their welfare payments. Fifteen women are randomly selected from the group of women who are about to begin the training program. Their average welfare payment $(\overline{Y}_1)$ is $160, and the sample standard deviation (s_1) is $50. One year after the training program, their average welfare payment $(\overline{Y}_2)$ is $145, and the sample standard deviation (s_2) is $45. The Pearson correlation between each subject's welfare payment before and after training (r_{12}) is .50.

Hypothesis The null hypothesis is that the population mean difference in welfare payments before and after the training program is zero. The one-tailed research hypothesis is that the population mean difference is greater than zero; that is, welfare payments are expected to be greater before the training programs than afterward.

Procedure

1. Calculate the test statistic:

$$t = \frac{(160-145)-0}{\sqrt{\dfrac{50^2+45^2-2(.50)(50)(45)}{15}}}$$

$$= 1.218.$$

2. Calculate the degrees of freedom: $df = 15 - 1 = 14$.
3. Given a level of significance of .05, a one-tailed research hypothesis, and 14 degrees of freedom, use Table C in the Appendix to make the statistical decision. Because the observed t of 1.218 is less than the table value of 1.761, accept the null hypothesis. This would be reported as $t_{.05} = 1.218$, $p > .05$. The job training program did not result in a significant reduction of welfare payments to its trainees.

SPSS

The SPSS program for this test is called T-TEST for paired samples (see Chapter 11 in the *SPSS Primer*). This program is limited to related samples from before-and-after designs in which the subjects serve as their own controls (method a).

A description of this procedure is based on the above example. Suppose that the variable name for the amount of welfare payments to the 15 women before job training is DOLLAR1; the variable name for amount of welfare payments 1 year after job training for these same 15 women is called DOLLAR2. The *t*-test for mean differences between related samples is then done in SPSS by the following statement:

```
1                      16
T-TEST                 PAIRS = DOLLAR1, DOLLAR2
```

The PAIRS specification indicates that the mean differences for each of the 15 pairs (or the difference between before and after welfare payments for each of the 15 subjects) on DOLLAR1 and DOLLAR2 will be calculated and tested for statistical significance.

Table 15.2 demonstrates what the SPSS printout would look like. The leftmost column, VARIABLE, gives the variable names for each of the means, DOLLAR1 and DOLLAR2. The next column, NUMBER OF CASES, reports the number of pairs used in the test. Here, it is 15 pairs. The MEAN column reports the mean ($\overline{Y}_1$ and $\overline{Y}_2$) for each variable: the mean for DOLLAR1 is 160.0000 and the mean for DOLLAR2 is 145.0000. The STANDARD DEVIATION column reports the estimated population standard deviation, s_1^2 and s_2^2, for each variable: for DOLLAR1 it is 50.000 and for DOLLAR2 it is 45.000. The STANDARD ERROR column reports the estimated population standard error of the mean, $s_1/\sqrt{n_1}$ and $s_2/\sqrt{n_2}$, for each variable: for DOLLAR1 it is 12.199 and for DOLLAR2 it is 11.628. The (DIFFERENCE) MEAN column reports the difference between the mean of DOLLAR1 and DOLLAR2. Thus, $160 - 145$ results in a mean difference of 15 dollars in this sample. The next column, STANDARD DEVIATION, refers to the sample or estimated population standard deviation of the mean differences (s_d) for the 15 pairs, which is 47.66. The STANDARD ERROR column reports the estimated population standard error of mean differences ($s_d/\sqrt{n}$) or denominator of the *t*-test, 12.316. The CORR column reports the Pearson *r* between DOLLAR1 and DOLLAR2, which is .50, and which has a 2-TAIL PROB. (probability) of .0478 and thus is significant at the .05 level. T VALUE is the value of the *t* statistic (1.22). The DEGREES OF FREEDOM to use the sampling distribution are 14, and the 2-TAIL PROB. associated with a *t* statistic of this size with 14 degrees of freedom is .3156. Because .3156 is greater than .05, a two-tailed test would lead to the decision to accept the null hypothesis. If a one-tailed test were done, the probability of .1578 (.3156/2) would indicate that the null hypothesis should be accepted—as it was in the above example—because .1578 is greater than the .05 level of significance.

Comment

If method b as described under "Purpose" is used to match pairs of subjects, there are two concerns to be cautious about. First, unless the

Table 15.2 An example of the printout for a t-test for related samples

VARIABLE	NUMBER OF CASES	MEAN	STANDARD DEVIATION	STANDARD ERROR	(DIFFERENCE) MEAN	STANDARD DEVIATION	STANDARD ERROR	CORR.	2-TAIL PROB.	T VALUE	DEGREES OF FREEDOM	2-TAIL PROB.
DOLLAR 1	15	160.0000	50.000	12.199								
					15.0000	47.660	12.316	.50	.0478	1.22	14	.3156
DOLLAR 2		145.0000	45.000	11.628								

characteristics used to match subjects into pairs result in a relatively high and positive relationship (r) between the means, matching may be less efficient than the comparison of unmatched random groups where the degrees of freedom would be twice as high; that is, the degrees of freedom would then refer to the number of subjects rather than the number of pairs. Second, the relationship (r) between means should be positive rather than negative if the standard error of mean differences is to be reduced by the matching process. Caution must be exercised, therefore, in choosing what subject characteristics will be used in the matching process. Overall, these two concerns suggest that matching should only be used as a last resort rather than a first preference for a research design.

15.6 ONE-WAY ANALYSIS OF VARIANCE

Purpose

This test determines statistically whether three or more population means (μ_1, μ_2, μ_3,...) estimated from their sample means (Y_1, Y_2, Y_3...) are equal. These means are from independent samples or groups. One-way analysis of variance is an extension of the t-test for comparing two means from two independent groups (Section 15.4) to comparing means from three or more independent groups. It is called "one-way" because there is only one independent variable whose values define each of the groups.

Null Hypothesis

The population means of all the independent groups are equal; in other words, they are all equal to a common population mean (μ).

$$H_0: \mu_1 = \mu_2 = \mu_3 = \ldots = \mu.$$

That is, under the null hypothesis, all the samples are conceived as being drawn from the same population, which has a population mean (μ).

Research Hypothesis

The population means of the independent groups are not equal. At least one group was drawn from a different population where the population mean was not μ.

The distinction between a one-tailed and a two-tailed test does not apply here. The research hypothesis is more similar to a two-tailed test, because it does not specify which mean will differ. However, the statistical test (F) itself is constructed as a one-tailed test. The reason for this will be explained below.

Statistical Test of The Null Hypothesis

$$F = \frac{MS_b}{MS_w} = \frac{\overset{k}{\underset{1}{\Sigma}} n_k (\overline{Y}_k - \overline{Y})^2 / (k-1)}{\overset{k}{\underset{1}{\Sigma}} \overset{n_k}{\underset{1}{\Sigma}} (Y - \overline{Y}_k)^2 / (n-k)}$$

where

F = an F test which as formulated is a random variable with an F distribution. F is the ratio of two estimated population variances, s_b^2 and s_w^2, or MS_b and MS_w, which are assumed to be equal to a common population variance σ ($\sigma_b^2 = \sigma_w^2 = \sigma$) under the null hypothesis. That is, if the data provide independent estimates (s_b^2 and s_w^2) of the same common population variance (σ), the population means will be equal. Under the null hypothesis, therefore, the F statistic will be equal to 1 (any number divided by itself is equal to 1) or close to one given random sampling fluctuations. If the population means are unequal, however, then the population variance (σ_b^2), which is estimated by (MS_b), will be greater than the population variance (σ_w^2), which is estimated by (MS_w). The F statistic will then be greater than 1. The F table will indicate whether this departure from 1 represents a significant difference between means.

To understand the meaning of σ_b^2 and its estimate MS_b, and the meaning of σ_w^2 and its estimate MS_w, a general discussion about variance is required. First, however, review Section 8.6.

An alternative name for variance is "mean square," abbreviated MS. A variance or mean square is the average of the sum of squared deviations about the mean. For an estimated population variance or mean square of a common variance, the defining formula is:

$$s^2 = MS = \frac{\Sigma(Y - \overline{Y})^2}{n-1} = \frac{TSS}{n-1}.$$

The variance numerator is the total sum of squared deviations about the mean (TSS), which may be partitioned into two additive or independent components:

$$TSS = SS_b + SS_w.$$

That is, the total sum of squares (TSS) is equal to the between sum of squares (SS_b) plus the within sum of squares (SS_w). This partitioning of TSS into two additive components is the reason this procedure is called the analysis of variance.

The between sum of squares (SS_b) is the sum of squared deviations of each group mean ($\overline{Y}_k$) about the total or overall mean ($\overline{Y}$). The subscript k labels the groups from 1 to the number of independent groups, and n_k is the sample size of each group:

$$SS_b = \overset{k}{\underset{1}{\Sigma}} n_k (\overline{Y}_k - \overline{Y})^2.$$

The within sum of squares (SS_w) is the sum of squared deviations of each subject's score (Y) from its group mean ($\overline{Y}_k$):

$$SS_w = \sum_1^k \sum_1^{n_k} (Y - Y_k)^2.$$

Now return to the definition of common variance. When the total sum of squares (TSS) is divided by its degrees of freedom ($n-1$), it defines the total variance or the mean square. These degrees of freedom may also be partitioned into two additive or independent components:

$$n - 1 = (k - 1) + (n - k).$$

The first set of degrees of freedom ($k-1$) is the degrees of freedom for SS_b. The second set of degrees of freedom ($n-k$) is the degrees of freedom for SS_w. If each sum of squares is divided by its degrees of freedom, the result is an estimate of a population variance or mean square. The sum of squares between groups (SS_b) may be divided by ($k-1$) to obtain the between mean square (MS_b). The sum of squares within groups (SS_w) may be divided by ($n-k$) to obtain the within mean square (MS_w). Both of these mean squares, under the null hypothesis, independently estimate a common population variance or mean square (σ).

$MS_b =$ the between mean square which estimates the population variance (σ_b^2). This estimate reflects variation of the independent sample means of each group from the total mean. If the data consist of independent and randomly selected samples with equal population means (as stated in the null hypothesis), MS_b provides an unbiased estimate of a common population variance (σ). That is, in the population, σ_b^2 will be equal to σ^2. However, MS_b will be a biased estimate if the population means are unequal (as stated in the research hypothesis), and thus MS_b will be larger than expected. If so, the F statistic will be significantly greater than 1.

$MS_w =$ the within mean square which estimates a population variance (σ_w^2). This estimate reflects variation of subjects' scores about their groups' independent sample mean. MS_w provides an unbiased estimate of a common population variance (σ) whether or not the null hypothesis is accepted. That is, in the population, σ_w^2 will equal σ.

In sum, the null hypothesis tested in one-way analysis of variance is that there is no significant difference between the population means. The hypothesis that all group means are equal is the same as saying that the variance of the means (estimated by the mean square between) is zero. The alternative research hypothesis states that the mean square between is significantly larger than the mean square within; that is, the means are not all equal and thus the variance of the means is greater than zero. A one-tailed test becomes necessary because the F test always has the potentially larger mean square (the mean square between) in the numerator and the smaller mean square (mean square within) in the denominator.

Sampling Distribution of This Test Statistic

F distribution.

Degrees of Freedom to Use This Sampling Distribution

df for $MS_b = k - 1$ This is the degrees of freedom for the potentially larger variance estimate placed in the numerator.

df for $MS_w = n - k$ This is the degrees of freedom for the smaller variance estimate placed in the denominator.

Assumptions Necessary to Use This Sampling Distribution

1. Simple random sampling is used to select all subjects.
2. The dependent variable, or the variable used for the calculation of means, is an interval-level measure.
3. For all of the independent samples or groups (defined by the independent variable), the dependent variable is normally distributed in the population.
4. For each of the independent samples or groups, the dependent variable has an equal population variance: $\sigma_1^2 = \sigma_2^2 = \sigma_3^2 = \ldots = \sigma_n^2$. This is called the homogeneity of variance or homoscedasticity assumption.

Decision

Given a level of significance, the research hypothesis, and the degrees of freedom, use Table F in the Appendix to make the statistical decision. If the observed F statistic is less than the table value, accept the null hypothesis. If the observed F statistic is equal to or greater than the table value, then there is support for the research hypothesis.

Example

Problem   In the detention home study, the 30 girls spent a different number of days in the detention home. One explanation for these different lengths of stay might be the seriousness of offense each girl was alleged to have committed. Seriousness of offense, an ordinal-level variable, has three score intervals—low, moderate, and high—each of which is used to define an independent sample of the group.

Question   Does seriousness of offense help explain the varying lengths of stay in the detention home?

Hypotheses   The null hypothesis is that there is no population mean difference on days spent in the detention home for the three independent groups: $\mu_1 = \mu_2 = \mu_3$. The research hypothesis is that not all population means are equal.

Procedure

1. Use the data in Table 4.1 to calculate the means and sum of squares for each group. The mean on days spent for the 30 girls is 7.1667.

Group 1:   Low seriousness

$n_1 = 12$

$\overline{Y} = 6$ days

$n_1(\overline{Y}_1 - \overline{Y})^2 = 12(6 - 7.1667)^2 = 16.33427$

$\Sigma(Y - \overline{Y}_1)^2 = (15 - 6)^2 + (8 - 6)^2 + (5 - 6)^2$
$\qquad\qquad\quad + (7 - 6)^2 + (4 - 6)^2 + (1 - 6)^2$
$\qquad\qquad\quad + (6 - 6)^2 + (9 - 6)^2 + (7 - 6)^2$
$\qquad\qquad\quad + (9 - 6)^2 + (6 - 6)^2 + (5 - 6)^2$
$\qquad\qquad = 56.$

Group 2:   Moderate seriousness

$n_2 = 9$

$\overline{Y}_2 = 6.7778$ days

$n_2(\overline{Y}_1 - \overline{Y})^2 = 9(6.7778 - 7.1667)^2 = 1.36119$

$\overset{n_2}{\Sigma}(Y - \overline{Y}_1)^2 = (12 - 6.7778) + \ldots + (4 - 6.7778)^2$
$\qquad\qquad = 71.55553.$

Group 3:   High seriousness

$n_3 = 9$

$\overline{Y}_3 = 9.1111$ days

$n_3(\overline{Y}_3 - \overline{Y})^2 = 9(9.1111 - 7.1667)^2 = 34.02672$

$\overset{n_3}{\Sigma}(Y - \overline{Y}_3)^2 = (11 - 9.1111)^2 + \ldots + (8 - 9.1111)^2$
$\qquad\qquad = 124.88886.$

2. Calculate the between sum of squares (SS_b) and the within sum of squares (SS_w).

$$SS_b = \overset{k}{\Sigma} n_k (\overline{Y}_k - Y)^2 = 16.33427 + 1.36119 + 34.02672$$
$$= 51.7222$$
$$SS_w = \overset{k}{\Sigma} \overset{n_k}{\Sigma} (Y - \overline{Y}_k)^2 = 56 + 71.55553 + 124.88887$$
$$= 252.4444.$$

3. Calculate the degrees of freedom:

$$df_b = k - 1 = 3 - 1 = 2$$
$$df_w = n - k = 30 - 3 = 27.$$

4. Calculate the between mean square (MS_b) and the within mean square (MS_w):

$$MS_b = SS_b / k - 1 = 51.7222 / 2 = 25.8611$$
$$MS_w = SS_w / n - k = 252.444 / 27 = 9.3498.$$

5. Calculate the F test:

$$F = \frac{MS_b}{MS_w} = \frac{25.8611}{9.3498} = 2.766.$$

6. Given a level of significance of .05, the research hypothesis, and 2 degrees of freedom for the greater mean square and 27 degrees of

freedom for the lesser mean square, use Table F in the Appendix to make the statistical decision. Because the observed F of 2.766 is less than the table value of 3.34, accept the null hypothesis. This would be reported as $F_{2,27} = 2.7666$, $p > .05$. Seriousness of offense does not help explain the number of days spent in the detention home. That is, the means for each group estimate an equal or common population mean and, thus, do not differ significantly.

SPSS

The SPSS program for one-way analysis of variance is not in the *SPSS Primer*, but is in the second edition of *SPSS*. The SPSS procedure for doing a one-way analysis of variance on the above data is:

```
1               16
ONEWAY          DAYS BY OFFENSE (1,3)
STATISTICS      1,3
```

The name ONEWAY, starting in column 1, calls forth the one-way analysis of variance program. The name of the dependent variable, DAYS, is typed in by starting at column 16. The word BY is followed by the name of the independent variable, OFFENSE. The numbers inside the parentheses indicate the lowest and the highest score intervals of the independent variable. Here the lowest score is 1 and the highest score for OFFENSE is 3. The STATISTICS card then specifies what statistics should be calculated.

Table 15.3 shows an example of the printout for this example. In the upper leftmost corner, the name of the dependent variable is shown: VARIABLE:DAYS. Underneath, an analysis of variance table is shown. The SOURCE column labels the three rows of this table: BETWEEN GROUPS, WITHIN GROUPS, and TOTAL. The next column provides the degrees of freedom (D.F.) for each row: 2 for between groups, 27 for with groups, and 29 for the total. The SUM OF SQUARES column gives the between sum of squares (51.7222), the within sum of squares (252.4444), and the total sum of squares (304.1667), which is the sum of the between and within sum of squares.

The MEAN SQUARES column gives the between mean square (25.8611) and the within mean square (9.3498). The F-RATIO column gives the value of the observed F test. The last column, F-PROB., provides the probability of observing an F of this size under the null hypothesis. Because .0808 is greater than the .05 level of significance, the null hypothesis should be accepted.

The statistics requested by the STATISTICS card are printed under the analysis of variance table. For each group (GROUP01, GROUP02,

Table 15.3. A printout for one-way analysis of variance

VARIABLE: DAYS

ANALYSIS OF VARIANCE

SOURCE	D.F.	SUM OF SQUARES	MEAN SQUARES	F-RATIO	F-PROB.
BETWEEN GROUPS	2	51.7222	25.8611	2.766	0.0808
WITHIN GROUPS	27	252.4444	9.3498		
TOTAL	29	304.1667			

GROUP	COUNT	MEAN	STANDARD DEVIATION	STANDARD ERROR	MINIMUM	MAXIMUM	95% CONF INT FOR MEAN
GRP01	12	6.0000	2.2563	0.6513	1.0000	9.0000	4.5664 TO 7.4336
GRP02	9	6.7778	2.9907	0.9969	3.0000	12.0000	4.4789 TO 9.0767
GRP03	9	9.1111	3.9511	1.3170	2.0000	14.0000	6.0740 TO 12.1482

TEST FOR HOMOGENEITY OF VARIANCES
COCHRAN'S C = MAX. VARIANCE/SUM (VARIANCES) = 0.5266, P = 0.213 (APPROX.)
BARTLETT – BOX F = 1.400, P = 0.247
MAXIMUM VARIANCE/MINIMUM VARIANCE = 3.066

GROUP03, and the TOTAL), the COUNT shows the sample size, the MEAN shows the sample mean of days spent, the STANDARD DEVIA-TION shows the estimated population standard deviation or days spent, the STANDARD ERROR gives the estimated population standard error or standard deviation of means (s_y), the MINIMUM shows the lowest score of days spent, and the MAXIMUM shows the highest score of days spent in that group. The last column, 95% CONF INT FOR MEAN, shows the 95 percent confidence interval for the mean of each group. A confidence interval for means is an interval around the sample mean. If a very large number of samples were drawn and such confidence intervals were placed around each sample mean, 95 percent of these intervals would contain the true population mean. Thus in GROUP01, for example, there is a 95 percent chance that the true population mean for days spent (for girls with a low seriousness of offense) falls in this interval (between 4.5664 days and 7.4336 days). The last set of statistics printed is TESTS FOR HOMOGENEITY OF VARIANCES. One of the assumptions necessary to use the F sampling distribution for the F test is that the population variances of each group are equal, that is, homogeneous or homoscedastic. Although this textbook does not include a discussion of tests for this assumption, it should be noted that SPSS provides such statistics: the Cochran C test, the Bartlett-Box F test, and the F_{max} test (maximum variance/minimum variance). The probabilities *(p)* reported here for the first two tests indicate that this assumption is met in this ex-ample. One reason for not discussing these tests further is that the F test in one-way analysis of variance is *robust;* that is, even if this assumption were not met, the statistical decision concerning F and mean differences is unlikely to change. Another reason is that these variance tests themselves are not robust—departures from normality in the population will affect the statistical decision. Thus, use of these tests in conjunction with analysis of variance is generally not recommended because of their limited practical utility.

Comments

If the F test indicates that there is a significant difference among the means, *then and only then* is it legitimate to do additional analyses to discover which group means differ significantly. A discussion of such analyses, called "multiple comparison tests for means," is beyond the scope of this textbook. However, a preliminary exploration may be done by using the *t*-test for comparing means from two independent samples (Section 15.4).

In analysis of variance, a distinction is made between a fixed effects model and a random effects model. A fixed effects model assumes that all groups of interest (scores of the independent variable) are included in the model. A random effects model assumes that groups have been randomly selected to represent some larger population of groups. Almost all applied research in the human services uses a fixed effects model; therefore, all remarks in this section apply only for the fixed effects model.

If there are only two groups defined by the independent variable, it can be shown that:

$$t_{n-2} = \sqrt{F_{1, n-k}}$$

That is, the critical value of a *two-tailed t-test* with $(n-2)$ degrees of freedom equals the square root of an F test with $(1, n-k)$ degrees of freedom. This direct parallel between the t-test and F test holds only for the case of two groups.

15.7   GOODNESS-OF-FIT CHI-SQUARE TEST FOR VARIANCE

Purpose

This test determines statistically whether a population variance (σ_1) estimated from an observed sample variance (s_1^2) is equal to a specified population variance (σ^2). That is, the estimated population variance could have been obtained in a random sample from the same population with the specified population variance. This is, therefore, a goodness-of-fit test. A specified population variance is a variance that is either previously known from prior research or that is based on a theoretical assumption about what population variance should be expected.

Null Hypothesis

The population variance (σ_1^2) of the sample is equal to (or comes from the same population as) the specified population variance (σ^2).

$$H_0: \sigma_1^2 = \sigma^2.$$

Research Hypothesis

1. The population variance of the sample is different from (or comes from a different population than) the specified population variance (a two-tailed test)

$$H_R: \sigma_1^2 \neq \sigma^2$$
or

2. The population variance of the sample is greater than the specified population variance (a one-tailed test)

$$H_R: \sigma_1^2 > \sigma^2$$
or

3. The population variance of the sample is less than the specified population variance (a one-tailed test)

$$H_R: \sigma_1^2 < \sigma^2.$$

Statistical Test of the Null Hypothesis

$$\chi^2 = \frac{(n-1)s_1^2}{\sigma^2}$$

where

$\chi^2 =$ a chi-square test statistic that as formulated is a random variable with a chi-square distribution
$n =$ the sample size
$s_1^2 =$ the estimated population variance based on the sample variance
$\sigma^2 =$ the known or specified population variance.

Sampling Distribution of This Test Statistic

Chi-square distribution.

Degrees of Freedom to Use This Sampling Distribution

$df = (n-1)$.

Assumptions Necessary to Use This Sampling Distribution

1. Simple random sampling is used to select subjects.
2. The variable is an interval-level measure.
3. The population distribution of this variable is normally distributed.

Decision

Use Table D in the Appendix with the appropriate degrees of freedom to make the statistical decision. If the research hypothesis is two-tailed as in hypothesis 1 where $\sigma_1^2 \neq \sigma^2$, then the probability column for finding the table value of the lower tail of the chi-square distribution should be either .975 ($\alpha = .05$) or .995 ($\alpha = .01$) and that of the upper tail should be either .025 ($\alpha = .05$) or .005 ($\alpha = .01$). Support for the research hypothesis is indicated by an observed χ^2 statistic falling *below* the table value for the lower tail or *above* the table value for the upper tail. If the research hypothesis is one-tailed as in hypothesis 2 where $\sigma_1^2 > \sigma^2$, then the probability column for finding the table value of the chi-square distribution should be either .05 ($\alpha = .05$) or .01 ($\alpha = .01$). Support for the research hypothesis is indicated by an observed chi-square statistic equal to or greater than the table value for the upper tail. If the research hypothesis is one-tailed as in hypothesis 3 where $\sigma_1^2 < \sigma^2$, then the probability column for finding the table value of the chi-square distribution should be either .950

$(\alpha = .05)$ or .990 $(\alpha = .01)$. Support for the research hypothesis is indicated by an observed chi-square statistic equal to or *below* the table value for the lower tail.

Example

Problem There is some evidence that salaries for nurses are more homogeneous or less variable than for doctors. Assume that the standard deviation of salaries for doctors is known to be $15,000 in the United States, but that similar information on nurses is unavailable. A random sample of 31 nurses is drawn, and the estimated population standard deviation in salaries is $5,000.

Question Is there greater homogeneity in salaries for nurses than for doctors?

Hypotheses The null hypothesis is that there is no difference in the homogeneity (that is, the population variances) of salaries. The research hypothesis is that nurses have greater homogeneity in salaries than doctors, and thus the variance of nurses' salaries should be smaller than the variance for doctors' salaries in the population.

Procedure

1. Calculate the variances on the basis of the given standard deviations.

    s^2 for nurses $= (5,000)^2 = 25,000,000$
    σ_1^2 for doctors $= (15,000)^2 = 225,000,000$.

2. Calculate the test statistic:

$$\chi^2 = \frac{(31-1)\,25,000,000}{225,000,000} = 3.333.$$

3. Calculate the degrees of freedom: $df = 31 - 1 = 30$.
4. Given a level of significance at .05, a one-tailed research hypothesis, and 30 degrees of freedom, use Table D in the Appendix to make the statistical decision. Since the observed statistic of 3.333 is less than the table value of 18.493, there is support for the research hypothesis. This would be reported as $\chi_{30}^2 = 3.333, p < .05$.

SPSS

Because SPSS does not provide a program, this test statistic must be calculated by hand.

15.8 *F* TEST FOR COMPARING TWO POPULATION VARIANCES

Purpose

This test determines statistically whether two population variances, both of which are estimated from sample information, are equal. The test compares two estimated population variances, s_1^2 and s_2^2, to determine whether they estimate a common population variance (σ^2) or whether they estimate the variance of two different populations, σ_1^2 and σ_2^2. Each sample is selected independently from a population.

Null Hypothesis

The population variances of the two samples are equal; that is, each estimated population variance (s_1^2 and s_2^2) estimates the same or common population variance (σ^2):

$$H_0: \sigma_1^2 = \sigma_2^2.$$

Research Hypothesis

1. The population variances of two samples are not equal; that is, each estimated population variance (s_1^2 and s_2^2) estimates a different population variance (σ_1^2 and σ_2^2). This is a two-tailed test

$$H_R: \sigma_1^2 \neq \sigma_2^2$$
or

2. The population variance of the first sample is greater than the population variance of the second sample (a one-tailed test)

$$H_R: \sigma_1^2 > \sigma_2^2$$
or

3. The population variance of the first sample is less than the population variance of the second sample (a one-tailed test)

$$H_R: \sigma_1^2 < \sigma_2^2.$$

Statistical Test of the Null Hypothesis

$$F = \frac{s_{\text{larger}}^2}{s_{\text{smaller}}^2}$$

where

$F =$ an F test statistic which as formulated is a random variable with an F distribution. If both estimated population variances estimate a common population variance, F will be 1 or close to 1 (the same number divided by itself will always equal 1). If they estimate different population variances, then F will be greater than 1 because the larger variance estimate is placed in the numerator. The larger estimate is in the numerator only to make tables of the F distribution shorter to print.

$s_{\text{larger}}^2 =$ the larger of the two estimated population variances, which may be either s_1^2 or s_2^2. If the research hypothesis 2 is being tested, then it is assumed that s_1^2 will be greater than s_2^2 and thus placed in the numerator. If s_1^2 is not larger, then it is obvious that this research hypothesis cannot be supported. If the research hypothesis 3 is being tested, then it is assumed that s_1^2 will be less than s_2^2 and thus s_2^2 will be placed in the numerator. If s_2^2 is not larger, then it is obvious that this research hypothesis cannot be supported.

$s_{smaller}^2$ = the smaller of the two estimated population variances, which may be either s_1^2 or s_2^2.

Sampling Distribution of This Test Statistic

F distribution.

Degrees of Freedom to Use This Sampling Distribution

$df_{larger} = n - 1$, where n is the sample size of the greater estimated population variance or the greater mean square (see Section 8.3 to understand why the variance is called a mean square).

$df_{smaller} = n - 1$, where n is the sample size of the smaller estimated population variance or lesser mean square.

Assumptions Necessary to Use This Sampling Distribution

1. Simple random sampling is used to select subjects from two independent groups.
2. The variable is an interval-level measure.
3. The population distribution of this variable is normally distributed.

Decision

Given a significance level and the type of research hypothesis, use Table F in the Appendix with the appropriate degrees of freedom to make the statistical decision. This F table is for one-tailed tests at the .05 or .01 levels of significance. For one-tailed tests (research hypothesis options 2 and 3), if the observed F is less than the table value, accept the null hypothesis. If the observed F is equal to or greater than the table value, then there is support for the research hypothesis. For a two-tailed test (research hypothesis 1), the only two-tailed F values that can be found from this table are for the .10 and .02 levels of significance. The .10 level of significance for a two-tailed test corresponds to the .05 level of significance for a one-tailed test; the .02 level of significance for a two-tailed test corresponds to the .01 level of significance for a one-tailed test. The upper limit of the rejection region is found by looking for an F value with $(df_{larger}, df_{smaller})$. The lower limit of the rejection region is found by: a) reversing the degrees of freedom such that you look for an F value of $(df_{smaller}, df_{larger})$, and b) taking the reciprocal of the table F value:

$$\frac{1}{F_{df_{smaller}, df_{larger}}}.$$

If the observed F statistic is greater than the table upper limit or smaller than the calculated lower limit of the rejection region, then there is support for the two-tailed research hypothesis.

206 Inferential Statistics

Example

Problem   Private vendors of group home services are to be reimbursed for costs under purchase of service contracts with the state. The eastern and western divisions of the state had the same average number of days to wait for reimbursement, but vendors in the eastern division complained that they had problems in getting their reimbursements in a timely fashion and thus were suffering difficulties in their cash flow. One possible explanation for their complaints was that the variance in days to reimbursement was greater for the eastern division than for the western division. Thus, even though both divisions had the same average number of days to wait, the eastern division may experience a greater spread (from much shorter to much longer) in number of days to reimbursement. To test this possibility, a random sample of 40 vendors was selected. There were 25 vendors from the eastern division in this sample and 15 vendors from the western division. The variance in number of days to reimbursement for the eastern division was $s_1^2 = 64$ and for the western division was $s_2^2 = 25$.

Question   Does the eastern division have a greater variance than the western division in the number of days to wait for their costs to be reimbursed?

Hypotheses   The null hypothesis is that there is no difference between the population variances for each division; that is, the population variance is the same for each division. The one-tailed research hypothesis is that the population variance for the eastern division is greater than the population variance for the western division.

Procedure

1. Calculate the F statistic. Because s_1^2 is larger than s_2^2, s_1^2 should be placed in the numerator.

$$F = \frac{64}{25} = 2.56.$$

2. Calculate the degrees of freedom:

$$df_{larger} = 25 - 1 = 24$$
$$df_{smaller} = 15 - 1 = 14.$$

3. Given a level of significance of .05, a one-tailed research hypothesis, and (24,14) degrees of freedom, use Table F in the Appendix to make the statistical decison. Because the observed F statistic of 2.56 is greater than the table value of 2.35, there is support for the research hypothesis. This would be reported as $F_{24,14} = 2.56$, $p < .05$. The eastern division does have greater variance in the number of days to wait for reimbursement.

Example

Problem In the detention home study described in Part I, the problem is to determine whether the variability of days spent in the detention home differs for two groups. One group is defined as 12 girls who have a low score on the seriousness of offense variable; the other group is defined as 18 girls who have moderate or high scores on the seriousness of offense variable. The sample standard deviation for the low seriousness of offense group on days spent is 2.256 days. The sample standard deviation for the higher seriousness of offense group on days spent is 3.605 days.

Question Which group has a greater variability for days spent in the detention home?

Hypotheses The null hypothesis is that there is no difference in the population variances for each group. The two-tailed research hypothesis is that the groups differ in terms of their population variances on days spent, but it remains unspecified as to which group has the greater variance.

Procedure

1. Square each sample standard deviation to provide an estimate of the population variance:

   low seriousness group: $s^2 = (2.256)^2 = 5.089$
   higher seriousness group: $s^2 = (3.605)^2 = 12.996$.

2. Calculate the F statistic, placing the larger estimated population variance in the numerator:

$$F = \frac{12.996}{5.089} = 2.55.$$

3. Calculate the degrees of freedom:

   low seriousness group: $df_{smaller} = 12 - 1 = 11$
   higher seriousness group: $df_{larger} = 18 - 1 = 17$.

4. Use Table F in the Appendix to determine the upper and lower limits of the rejection region using a .10 level of significance with 17 and 11 degrees of freedom.

   upper rejection limit: $F_{17,11} = 2.70$

Note that there is no exact value given for the larger degrees of freedom of 17. When this happens, it is easiest to use the next lowest degrees of freedom. Thus, 2.70 really represents an F with 16 and 11 degrees of freedom.

   lower rejection limit: $\dfrac{1}{F_{11,17}} = \dfrac{1}{2.41} = .41$

Because the observed F statistic of 2.55 is less than the upper limit (2.70) and greater than the lower limit (.41), accept the null hypothesis. There is no significant difference between the groups' population variances.

SPSS

The SPSS program for this F test is given in the T-TEST program discussed in Section 15.4. Look at Table 15.1 in that section. The column labeled STANDARD DEVIATION provides the sample standard deviation for each group. The column labeled F VALUE provides the value of the test statistic, which is 2.55. The column labeled 2-TAIL PROB. provides the two-tailed probability of observing an F statistic of this size. The probability given is .118, meaning that 12 times out of 100 random samples of this size would an F of this size be observed. In the above problem, a .10 level of significance was used to test a two-tailed hypothesis. Because the observed probability of .118 is greater than the present level of .10, the null hypothesis should be accepted. If the research hypothesis was *one-tailed*, the probability reported in SPSS should be divided in half, which, if greater than the preset level of significance, leads to the decision to accept the null hypothesis.

15.9 *t*-TEST FOR COMPARING
POPULATION VARIANCES FROM RELATED SAMPLES

Purpose

This test determines statistically whether two population variances (σ_1^2 and σ_2^2) estimated from sample information (s_1^2 and s_2^2) are equal when the subjects belong to related samples. (Related samples may be of two kinds, as described in Section 15.5.) The purpose of this test is essentially the same as described for the F test in Section 15.8, except that the use of related samples may reduce the sampling error due to differences in subject characteristics and thus increase the likelihood that a real difference in population variances between groups will not be obscured.

Null Hypothesis

The population variances for the related groups are equal, or the difference between population variances from related samples is zero.

$$H_0: \sigma_1^2 = \sigma_2^2 \quad \text{or} \quad \sigma_1^2 - \sigma_2^2 = 0.$$

Research Hypotheses

1. The population variances for the related groups are not equal (a two-tailed test)

$$H_R: \sigma_1^2 \neq \sigma_2^2$$

or

2. The population variance for group 1 is greater than the population variance for group 2 (a one-tailed test)

$$H_R: \sigma_1^2 > \sigma_2^2$$

or

3. The population variance for group 1 is less than the population variance for group 2

$$H_R: \sigma_1^2 < \sigma_2^2.$$

Statistical Test of the Null Hypothesis

$$t = \frac{(s_1^2 - s_2^2)\sqrt{n-2}}{2s_1 s_2\sqrt{1-r^2}}$$

where

t = a t-test which as formulated is a random variable with a t distribution. When there is little or no difference between s_1^2 and s_2^2 (that is, random sampling fluctuations could account for the differences), t will be equal or close to zero. As the difference between these two estimates of population variances increases, so will the size of t increase given a constant n and r.

s_1^2 = the estimated population variance for group 1, that is, the sample variance

s_1 = the sample standard deviation of group 1 which, when squared, will be the above

s_2^2 = the estimated population variance for group 2, that is, the sample variance

s_2 = the sample standard deviation of group 2 which, when squared, will be the above

n = the number of pairs of subjects (*not* the total number of subjects)

r^2 = the symbol for the square of Pearson r. The Pearson r is the correlation between the two groups on the variable tested for equal variances. The squared correlation indicates the extent to which one variable can be predicted from knowledge of the other, if there is a linear relationship between variables. The larger the correlation in this situation, the more likely it is that a significant difference between variances may be found. Note that in Section 15.8, this correlation is not part of the F test, and thus differences between variances there would have to be larger to find a significant difference.

Sampling Distribution of This Test Statistic

t distribution.

Degrees of Freedom to Use This Sampling Distribution

$n - 2$.

Assumptions Necessary to Use This Sampling Distribution

1. Simple random sampling of subjects for a before-and-after design, or random assignment of equal numbers of subjects to an experimental and control group.
2. The variable used for the calculation of the mean is an interval-level measure.
3. The population distribution of the variable is normally distributed in both groups.

Decision

Given a level of significance, the type of research hypothesis, and the degrees of freedom, use Table C in the Appendix to make the statistical decision. For the two-tailed research hypothesis 1, if the absolute value (see Section 2.7) of the observed t statistic is equal to or greater than the table value, then there is support for the research hypothesis. Otherwise, accept the null hypothesis. For the one-tailed research hypothesis 2, if s_1^2 is greater than s_2^2 and if the observed t statistic is equal to or greater than the table value, then there is support for the research hypothesis. Otherwise, accept the null hypothesis. For the one-tailed research hypothesis 3, if s_1^2 is less than s_2^2 and if the absolute value of the observed t statistic is equal to or greater than the table value, there is support for the research hypothesis. Otherwise, accept the null hypothesis.

Example

Problem   Once a year, new paraprofessional staff are given a staff training program to learn how to complete the necessary weekly reporting forms accurately. This year, 20 new staff members participated in the training program. A trial test on filling out the reporting forms was given before the training started. The sample standard deviation (s_1) was 12 errors. Two weeks after the training program, this test was given again to the same 20 people. The sample standard deviation (s_2) was 8 errors. The Pearson correlation between scores on the test before and after training was .60.

Question   Does training reduce the variability of errors in filling out record forms?

Hypotheses   The null hypothesis is that the population variance for errors is the same before and after the training program. The one-tailed research hypothesis is that the population variance before training is greater than the population variance after training.

Procedure

1. Calculate the t-test:

$$t = \frac{(144-64)\sqrt{20-2}}{2(12)(8)(\sqrt{1-.60^2})} = 2.76.$$

2. Calculate the degrees of freedom: $df = 20 - 2 = 18$.
3. Given a level of significance of .05, a one-tailed research hypothesis, and 18 degrees of freedom, use Table C in the Appendix to make the statistical decision. Because the observed t statistic of 2.762 is greater than the table value of 1.734, there is support for the research hypothesis. This would be reported as $t_{18} = 2.762$, $p < .05$.

SPSS

Because SPSS does not have a program for this test, it must be calculated by hand.

Comment

If the F test in Section 15.8 had been calculated instead, the same decision would have been made, but the observed F statistic of 2.55 was much closer to the upper limit of the rejection limit or critical value of 2.25. Thus, the related sample procedure is most useful when differences between groups are small and yet a significant difference is expected.

15.10 t-TEST FOR PEARSON r EQUAL TO ZERO IN THE POPULATION

Purpose

This test determines statistically whether there is a linear association (ρ) between two interval-level variables in the population. The population correlation is estimated by Pearson r.

Null Hypothesis

There is no linear relationship in the population.

$$H_0: \rho = 0.$$

Research Hypothesis

1. There is a linear relationship in the population (a two-tailed test)

$$H_R:$$
$$\rho \neq 0$$
or

2. There is a *positive* linear relationship in the population (a one-tailed test)

$$H_R:$$
$$\rho > 0$$
or

3. There is a *negative* linear relationship in the population (a one-tailed test)

$$H_R:$$
$$\rho < 0.$$

Statistical Test of the Null Hypothesis

$$t = r\sqrt{\frac{n-2}{1-r^2}}$$

where

  t = a Student's t-test which as formulated is a random variable with a t distribution. The t represents the number of standard errors from this distribution's mean of zero. The larger the observed Pearson r is, the greater the size of t becomes given a constant sample size.
  r = the observed Pearson correlation in the sample
  n = the sample size.

Sampling Distribution of This Test Statistic

t distribution.

Degrees of Freedom to Use This Sampling Distribution

$n-2$.

Assumptions Necessary to Use This Sampling Distribution

1. Simple random sampling is used to select subjects.
2. The variables are interval-level measures.
3. The relationship between variables is linear in the population.
4. Each variable is normally distributed in the population.
5. The joint distribution of both variables together is a bivariate normal distribution in the population: a) Y scores are normally distributed for each score of X, and b) X scores are normally distributed for each score of Y.
6. The joint distribution of both variables together is homoscedastic in the population: a) Y scores have the same variance around each score of X, and b) X scores have the same variance around each score of Y.

Decision

Given a level of significance, the type of research hypothesis, and the degrees of freedom, use Table C in the Appendix to make the statistical decision. For the two-tailed research hypothesis 1, if the absolute value (see Section 2.7) of the t statistic is equal to or greater than the table value, then there is support for the research hypothesis. Otherwise, accept the null hypothesis. For the one-tailed research hypothesis 2, if r is positive and the observed t is equal to or greater than the table value, then there is support for the research hypothesis. Otherwise, accept the null hypothesis. For a one-tailed research hypothesis 3, if r is negative and the absolute value of the observed t statistic is equal to or greater than the table value, then there is support for the research hypothesis. Otherwise, accept the null hypothesis.

Example

Problem   A random sample of 62 male parolees was selected from a state department of corrections. A Pearson correlation of $-.42$ was found between two variables: age and frequency of alcohol consumption as reported in a survey questionnaire.

Question   Is there an inverse negative relationship between age and frequency of alcohol consumption? That is, as age increases, does frequency of alcohol consumption decrease?

Hypotheses   The null hypothesis is that there is no linear relationship in the population. The one-tailed research hypothesis is that there is a negative, linear relationship between age and frequency of alcohol consumption in the population.

Procedure

1. Calculate the t-test:

$$t = -.42\sqrt{\frac{62-2}{1-(-.42)^2}}$$
$$= -8.535.$$

2. Calculate the degrees of freedom: $df = n - 2 = 60$.
3. Given the level of significance at .05, a one-tailed research hypothesis, and 60 degrees of freedom, use Table C in the Appendix to make the statistical decision. Because the absolute value of the observed statistic, 8.535, is greater than the table value of 1.671, this provides support for the research hypothesis. This would be reported as $t_{60} = -8.535, p < .05$. There is a moderate, and statistically significant, inverse linear relationship between age and frequency of alcohol consumption.

SPSS

The PEARSON CORR program calculates this test statistic and its probability of occurring (review section 11.6 and Table 11.5). The printout provides only the probability for a one-tailed hypothesis test; it does not provide the observed t statistic nor the degrees of freedom.

Comment

The hypothesis test for Pearson r (H_0: $\rho = 0$) provides the same result as the hypotheses tests for slope (H_0: $\beta = 0$) and R^2 (H_0: $R^2_{pop} = 0$) in simple linear regression. In other words, when there are *two* variables, the null hypothesis could be written:

$$H_0: \rho = \rho^2 = \beta = R^2_{pop} = 0.$$

The hypothesis that the correlation is zero in the population is the same as saying that the r^2, slope, and R^2 are zero in the population—but only when there are two variables.

15.11   FISHER Z TRANSFORMATION FOR PEARSON r TO COMPARE WITH A KNOWN BUT NONZERO POPULATION VALUE

Purpose

This test determines statistically whether an observed Pearson correlation (r) estimates a population correlation (ρ) that is equal to a previously known but nonzero population correlation (ρ_{known}). This is, therefore, a goodness-of-fit test.

Null Hypothesis

The estimated population Pearson correlation (ρ) is equal to (or comes from the same population as) the known population correlation (ρ_{known}).

$$H_0: \rho = \rho_{known}.$$

Research Hypothesis

1. The estimated population Pearson correlation is not equal to a previously known population correlation (a two-tailed test)

$$H_R: \rho \neq \rho_{known}$$

or

2. The estimated population Pearson correlation is *greater than* the previously known population correlation (a one-tailed test)

$$H_R: \rho > \rho_{known}$$

or

3. The estimated population Pearson correlation is *less than* the previously known population correlation (a one-tailed test)

$$H_R: \rho < \rho_{known}.$$

Statistical Test of the Null Hypothesis

$$z = (Z_r - Z_{known})(\sqrt{n-3})$$

where

$z =$ a random variable with a normal distribution. The z represents the number of standard errors from this distribution's mean of zero. The greater the discrepancy between r and ρ_{known}, the larger z becomes given a constant sample size.

$Z_r =$ the observed Pearson r from the sample transformed to Z_r by using Table E in the Appendix

$Z_{known} =$ the previously known population correlation (ρ_{known}) transformed to Z_{known} using Table E in the Appendix

$n =$ the sample size in which the sample Pearson r was observed.

Sampling Distribution of This Test Statistic

The normal distribution.

Degrees of Freedom to Use This Sampling Distribution

None, but degrees of freedom $(n-3)$ are included in the test statistic.

Assumptions Necessary to Use This Sampling Distribution

The same assumptions apply here as for Pearson r in Section 15.10.

Decision

Use Table C in the Appendix to make the statistical decision. For the two-tailed research hypothesis 1, if the absolute value (see Section 2.7) of the observed z statistic is equal to or greater than the table z value, then there is support for the research hypothesis. Otherwise, accept the null hypothesis. For the one-tailed research hypothesis 2, if r is greater than ρ_{known} and if the observed z statistic is equal to or greater than the table z value, then there is support for the research hypothesis. Otherwise, accept the null hypothesis. For the one-tailed research hypothesis 3, if r is less than ρ_{known} and if the absolute value of the observed z statistic is equal to or greater than the table z value, then there is support for the research hypothesis. Otherwise, accept the null hypothesis.

Example

Problem   In the past, staff satisfaction with supervision in community residences has been found to correlate .49 with satisfaction with pay. That is, satisfaction with supervision increases as satisfaction with pay increases. In a random sample of 103 such employees who had received a 10 percent raise in salary 6 months ago, the correlation between satisfaction with supervision and satisfaction with pay was found to be .60.

Question   Does the recently found sample correlation of .60 fit with previous knowledge that the true population correlation is .49?

Hypothesis   The null hypothesis is that the estimated population correlation of .60 comes from the same population where the known correlation is .49. In other words, the difference between the estimated and known population correlations is not so large as to suspect that the populations differ significantly in this respect. The one-tailed research hypothesis is that the population correlation should be greater for the sample (because the sample had received salary increases) than the known population correlation. That is, the observed correlation represents a correlation from a population that differs from the population in which the known population correlation was found.

Procedure

1.   Transform the Pearson correlations to Z scores using Table E in the Appendix:

$$\rho_{known} = .49 \qquad Z_{known} = .536$$
$$r = .60 \qquad Z = .693.$$

2. Calculate z:

$$z = (.693 - .536)(\sqrt{103 - 3})$$
$$= 1.57.$$

3. With a level of significance set at .05, use Table C in the Appendix to make the statistical decision. Because r is greater than ρ_{known} but the observed z statistic of 1.57 is less than 1.645, the null hypothesis is accepted. This would be reported as $z = 1.57$, $p > .05$.

15.12   FISHER Z TRANSFORMATION TEST FOR COMPARING PEARSON CORRELATIONS FROM DIFFERENT SAMPLES OR GROUPS BUT INVOLVING THE SAME TWO VARIABLES

Purpose

This test assesses the equality of two population Pearson correlations, using the same variables, observed in two independent samples or groups. The population correlations are estimated from sample Pearson r's.

Null Hypothesis

The population correlation (ρ_1) between two variables estimated in one sample is equal to the population correlation (ρ_2) between the same two variables estimated in a different independent sample or group.

$$H_0: \rho_1 = \rho_2.$$

Research Hypothesis

1. The two population correlations are unequal (a two-tailed test)

$$H_R: \rho_1 = \rho_2$$
or

2. One population correlation (ρ_1) is greater than the other population correlation (ρ_2); this is a one-tailed test

$$H_R: \rho_1 > \rho_2$$
or

3. One population correlation (ρ_1) is less than the other population correlation (ρ_2); this is a one-tailed test

$$H_R: \rho_1 < \rho_2.$$

Statistical Test for the Null Hypothesis

$$z = \frac{(Z_1 - Z_2)}{\sqrt{\dfrac{1}{n_1 - 3} + \dfrac{1}{n_2 - 3}}}$$

where

z = a random variable which as formulated has a standard normal distribution. z represents the number of standard errors from this distribution's mean of zero. The larger the difference between the transformed Pearson correlations, the greater the size of z becomes, given constant sample sizes.

$Z_1 = r_1$ transformed to a Z value (from Table E in the Appendix)

$Z_2 = r_2$ transformed to a Z value (from Table E in the Appendix)

n_1 = the sample size of the sample in which r_1 was observed

n_2 = the sample size of the sample in which r_2 was observed.

Sampling Distribution of This Test Statistic

The normal distribution.

Degrees of Freedom to Use This Sampling Distribution

None.

Assumptions Necessary to Use This Sampling Distribution

The same assumptions apply here as for Pearson r.

Decision

Given a level of significance and the type of research hypothesis, use Table C in the Appendix to make the statistical decision. For the two-tailed research hypothesis 1, if the absolute value (see Section 2.7) of the observed z statistic is equal to or greater than the table z value, then there is support for the research hypothesis. Otherwise, accept the null hypothesis. For the one-tailed research hypothesis 2, if r_1 is greater than r_2 and if the observed z statistic is equal to or greater than the table z value, then there is support for the research hypothesis. Otherwise, accept the null hypothesis. For the one-tailed research hypothesis 3, if r_1 is less than r_2 and if the absolute value of the observed z statistic is equal to or greater than the table z value, then there is support for the research hypothesis. Otherwise, accept the null hypothesis.

Example

Problem   A random sample of 103 institutionalized mental health patients and 53 community mental health center outpatients, who were previously institutionalized, was selected. For the institutionalized patients, the correlation between number of treatment days and cost of psychiatric care for each patient was − .45; that is, the number of treatment days increased as the cost of psychiatric care decreased. For the community outpatients, the correlation was .10; that is, as the number of treatment days increased so did the cost of psychiatric care.

Question   Is there a difference between the association of number of treatment days and cost of psychiatric care for the community outpatients and for the institutionalized patients?

Hypotheses   The null hypothesis is that both population correlations are equal. The two-tailed research hypothesis is that the community outpatients have a different population correlation between number of treatment days and the cost of psychiatric care than do the institutionalized patients.

Procedure

1. Transform the observed correlations to Z scores using Table E in the Appendix:

$$r \text{ (institutionalized)} = -.45 \qquad Z = -.485$$
$$r \text{ (community)} = .10 \qquad Z = .100.$$

2. Calculate z:

$$z = \frac{(-.485 - .100)}{\sqrt{\dfrac{1}{103 - 3} + \dfrac{1}{53 - 3}}} = -3.38.$$

3. With the level of significance at .05 and a two-tailed research hypothesis, use Table C in the Appendix to make the statistical decision. Because the absolute value of the z statistic is greater than the table value of 1.960, there is evidence to support the research hypothesis. This would be reported as $z = -3.38, p < .05$.

SPSS

This test must be calculated by hand, because SPSS does not provide a program.

15.13   *t*-TEST FOR COMPARING PEARSON CORRELATIONS THAT SHARE ONE VARIABLE IN COMMON IN THE SAME SAMPLE OR GROUP

Purpose

This test assesses the equality of two population Pearson correlations (ρ_{yx}, ρ_{yz}), using three variables, (X, Y, and Z). This comparison assumes, then, that there is one common variable between both correlations. Here, the common variable is Y. Both population correlations are estimated by Pearson r's observed in the same sample.

Null Hypothesis

The population correlation (ρ_{yx}) between the X and Y variables is equal to the population correlation (ρ_{yz}) between the Y and Z variables.

$$H_0: \rho_{yx} = \rho_{yz}.$$

Research Hypothesis

1. The two population correlations are unequal (two-tailed hypothesis)

$$H_R: \rho_{yx} \neq \rho_{yz}$$
or

2. The population correlation for X and Y is greater than for the population correlation for Y and Z (one-tailed hypothesis)

$$H_R: \rho_{yx} > \rho_{yz}$$
or

3. The population correlation for X and Y is less than the population correlation for Y and Z

$$H_R: \rho_{yx} < \rho_{yz}.$$

Statistical Test of the Null Hypothesis

$$t = (r_{yx} - r_{yz}) \sqrt{\frac{(n-3)(1 - r_{xz})}{2(1 - r_{yx}^2 - r_{xz}^2 - r_{yz}^2 + 2r_{yx}r_{xz}r_{yz})}}$$

where

t = a Student's t-test that is formulated as a random variable with a t distribution. The t represents the number of standard errors from this distribution's mean. The larger the difference between the observed Pearson correlations, the greater t becomes given a constant sample size.
r_{yx} = the Pearson correlation between variables X and Y
r_{yz} = the Pearson correlation between variables Y and Z
r_{xz} = the Pearson correlation between variables X and Z
n = the sample size.

Sampling Distribution of This Test Statistic

t distribution.

Degrees of Freedom to Use This Sampling Distribution

$n - 3$.

Assumptions Necessary to Use This Sampling Distribution

The same assumptions apply here as for Pearson r.

Decision

Given a level of significance, the type of research hypothesis, and the degrees of freedom, use Table C in the Appendix to make the statistical decision. For the two-tailed research hypothesis 1, if the absolute value (see Section 2.7) of the observed t statistic is equal to or greater than the table value, then there is support for the research hypothesis. Otherwise, accept the null hypothesis. For the one-tailed research hypothesis 2, if r_{yx} is greater than r_{yz} and if the observed t statistic is equal to or greater than the table value, then there is support for the research hypothesis. Otherwise, accept the null hypothesis. For the one-tailed research hypothesis 3, if r_{yx} is less than r_{yz} and if the absolute value of the observed t statistic is equal to or greater than the table value, then there is support for the research hypothesis. Otherwise, accept the null hypothesis.

Example

Problem   In a random sample of 43 men on public welfare, the number of months they had received welfare benefits was obtained (variable Y). Additional information was collected on their degree of physical disability (variable X) and their age (variable Z). The Pearson correlations were:

	X	Y	Z
X	100	.60	.30
Y		1.00	.40
Z			1.00

Question   Does degree of physical disability have a higher correlation with number of months receiving welfare benefits than does age?

Hypotheses   The null hypothesis is that number of months receiving welfare benefits correlates equally with degree of physical disability and with age. The one-tailed research hypothesis is that physical disability will have a higher correlation with number of months receiving welfare benefits than will age.

Procedure

1. Calculate t:

$$t = (.60 - .40)\sqrt{\frac{(43 - 3)(1 - .30)}{2(1 - .60^2 - .30^2 - .40^2 + 2(.60)(.30)(.40))}}$$
$$= 1.396.$$

2. Calculate the degrees of freedom: $df = n - 3 = 43 - 3 = 40$.
3. Given the significance level of .05 and a one-tailed research hypothesis, use Table C in the Appendix. Because the observed t of 1.396 is less than the table value of 1.684 with 40 degrees of freedom, the null hypothesis is accepted. This would be reported as $t_{40} = 1.396$, $p > .05$. Both variables, degree of physical disability and age, correlate equally with number of months receiving welfare benefits in the population.

SPSS

Because SPSS does not provide this test, it must be calculated by hand.

Appendix

Statistical Tables

222

Table A. A table of random numbers

Row	1	2	3	4	5	6	7	8	9	10	11	12	13	14	15	16	17	18	19	20	21	22	23	24	25	26	27	28	29	30	31	32	Row
1	2	7	8	9	4	0	7	2	3	2	5	4	2	6	7	1	6	8	5	9	1	3	5	4	0	3	6	6	7	6	5	1	1
2	2	2	6	0	4	1	7	1	3	8	7	3	6	7	9	4	2	1	3	8	9	0	3	4	9	0	2	6	3	0	9	8	2
3	9	9	6	6	3	9	4	9	1	0	5	1	5	2	2	7	5	2	5	3	4	1	3	9	5	8	1	3	8	2	9	2	3
4	7	7	5	5	9	2	7	5	7	8	0	8	8	5	0	6	0	5	9	0	5	7	4	5	2	0	6	1	6	4	2	0	4
5	4	4	3	6	6	3	9	8	2	1	7	9	7	6	4	2	4	9	6	0	3	6	3	5	3	9	9	1	8	5	1	3	5
6	8	2	0	8	8	7	7	6	0	2	2	3	1	1	1	6	4	8	5	2	2	3	4	2	2	6	5	2	2	4	9	6	6
7	0	8	7	5	3	3	6	4	2	6	8	3	1	6	5	0	0	5	5	7	8	1	0	1	2	9	1	4	3	4	7	6	7
8	9	0	1	9	0	8	4	6	6	8	6	3	3	2	2	3	7	4	7	5	1	5	7	6	3	7	9	4	5	5	3	5	8
9	5	4	0	6	7	4	0	0	0	1	9	5	9	9	1	8	1	4	7	4	9	8	7	2	4	3	0	8	6	4	2	7	9
10	1	9	5	4	1	5	2	6	2	9	4	1	1	5	8	4	4	4	6	1	8	7	8	6	4	8	7	4	4	0	5	8	10
11	5	6	2	4	4	8	7	2	8	3	6	1	5	9	8	6	2	2	9	1	9	0	4	8	1	0	1	3	5	3	4	4	11
12	7	9	2	5	1	9	7	9	3	1	8	6	8	7	7	6	6	5	0	3	8	1	1	2	4	7	8	9	1	7	5	2	12
13	3	3	3	5	9	5	1	0	0	8	2	5	6	3	5	4	6	5	7	2	6	7	8	9	9	9	8	0	9	1	5	3	13
14	1	9	3	4	0	0	9	9	5	7	4	1	5	9	4	7	6	4	8	2	6	4	4	1	8	8	1	5	4	3	8	0	14
15	5	4	4	2	2	3	3	7	9	1	0	9	6	2	9	7	4	7	6	1	1	6	1	2	2	9	5	8	4	4	8	6	15
16	2	9	8	5	2	5	9	3	2	0	4	9	0	6	4	4	2	1	5	7	3	6	5	5	4	5	7	9	6	6	4	0	16
17	9	7	6	6	3	5	0	3	3	3	1	7	5	0	9	6	9	3	3	9	2	1	1	0	0	1	3	7	7	3	7	3	17
18	5	5	2	4	3	3	0	8	5	3	5	7	5	8	3	5	9	8	4	5	4	6	3	9	2	7	1	1	4	9	1	9	18
19	4	1	4	9	4	3	6	6	2	9	7	4	6	2	5	6	9	0	3	6	1	4	0	3	5	9	7	1	8	0	6	9	19
20	1	1	9	8	8	8	6	6	7	0	9	7	9	6	9	9	4	0	6	0	0	5	9	6	5	1	4	2	0	4	1	9	20
21	6	9	9	8	3	3	3	5	9	6	6	7	7	6	0	4	5	3	4	5	7	3	0	6	1	0	3	0	0	3	5	0	21
22	7	0	0	3	8	1	5	4	7	9	5	2	6	9	9	7	3	2	5	0	2	3	5	3	9	7	4	8	9	4	1	5	22
23	3	7	0	2	1	5	5	6	9	0	1	7	8	9	6	6	6	0	7	8	1	9	6	7	4	8	9	6	3	6	5	1	23

Column number

24	2	1	5	5	4	2	3	4	0	7	4	8	2	2	7	1	3	9	2	3	0	6	5	6	0	5	6	0	0
25	2	1	3	6	4	8	9	0	7	7	1	8	3	4	0	6	9	4	7	4	2	8	2	4	9	4	8	7	0
26	8	6	2	1	7	6	1	7	1	4	4	4	5	7	4	2	2	4	6	4	3	2	2	5	7	7	1	3	3
27	3	0	7	0	7	5	6	6	4	4	3	1	6	7	9	8	1	2	7	4	6	9	0	7	5	8	4	1	3
28	2	7	2	0	9	7	7	8	2	3	2	9	5	2	5	5	3	0	1	3	2	6	3	0	8	3	7	1	2
29	9	0	1	3	2	3	5	8	8	0	9	8	0	7	0	1	4	7	8	2	9	3	2	7	9	0	2	2	0
30	6	6	3	8	1	2	5	9	4	3	1	2	7	5	8	8	5	4	0	0	8	2	8	9	6	5	9	9	6
31	9	5	7	5	6	7	1	6	5	5	5	1	2	1	6	6	3	3	8	1	7	9	3	9	5	6	2	9	2
32	0	4	4	5	1	4	5	0	4	1	8	7	3	9	3	1	9	5	9	0	5	5	4	1	1	9	9	7	7
33	1	0	9	6	1	8	2	2	9	5	6	8	3	1	6	4	4	1	9	0	7	7	3	4	3	4	4	2	8
34	2	3	0	8	1	4	5	0	9	6	0	0	1	6	2	5	1	3	9	6	8	5	0	2	6	1	1	3	0
35	1	7	8	8	6	2	9	8	5	3	0	9	1	3	0	6	5	1	4	0	5	7	0	9	8	6	6	8	4
36	0	6	8	6	7	6	2	2	7	7	4	3	3	8	5	0	7	2	4	6	7	6	5	6	7	8	8	1	7
37	9	9	8	3	2	4	7	3	8	6	2	2	2	1	0	3	0	0	3	2	6	1	1	8	1	1	3	8	4
38	8	3	3	4	4	4	4	8	3	3	7	6	0	8	4	2	5	8	7	1	9	7	8	0	6	0	6	2	8
39	4	9	5	6	8	2	1	0	2	3	0	4	4	7	0	3	6	7	1	8	1	2	8	2	8	9	5	3	0
40	2	1	3	7	8	7	6	1	2	4	1	1	8	7	0	4	8	0	2	8	9	6	6	0	9	0	0	6	8
41	6	0	4	0	0	4	5	3	8	6	8	1	1	5	2	4	2	7	4	3	8	8	7	4	9	9	9	5	6
42	9	3	5	8	5	0	1	3	5	0	9	1	2	9	8	6	1	8	4	0	5	1	9	3	2	9	9	5	0
43	3	6	3	9	7	7	2	8	7	8	1	6	0	6	6	7	4	7	7	0	4	5	9	3	3	5	5	1	8
44	4	1	8	9	1	9	0	0	4	3	4	1	0	8	1	5	7	9	8	1	4	7	8	2	5	0	0	4	3
45	7	8	4	2	1	7	4	1	0	6	6	5	3	6	9	1	5	2	8	0	5	6	5	3	1	4	4	7	2
46	1	0	1	0	9	4	1	9	5	6	3	6	1	0	4	2	7	2	6	2	4	2	2	8	7	5	0	3	4
47	6	8	7	7	6	3	0	1	8	6	7	4	5	6	2	9	6	7	5	2	0	0	5	8	6	0	4	3	4
48	8	5	7	8	7	3	6	3	1	1	6	3	5	3	0	0	5	1	4	8	5	8	0	2	0	4	5	6	6
49	7	7	1	6	2	9	0	0	1	1	8	9	2	2	9	3	9	9	7	2	1	1	0	2	4	5	8	3	5
50	0	2	5	4	6	3	3	3	5	9	8	5	2	4	0	0	0	9	3	6	6	7	9	5	8	8	4	4	4

This table is reprinted from John T. Roscoe, *Fundamental Research Statistics for the Behavioral Sciences*, Second Edition (New York, Holt, Rinehart and Winston, 1975, pp. 410–411).

Table B. Areas under the normal curve (proportion of total area under the normal curve between mean ordinate and ordinate at given z distance from the mean)

	Second decimal place in z									
	.00	.01	.02	.03	.04	.05	.06	.07	.08	.09
.0	.0000	.0040	.0080	.0120	.0160	.0199	.0239	.0279	.0319	.0359
.1	.0398	.0438	.0478	.0517	.0557	.0596	.0636	.0675	.0714	.0753
.2	.0793	.0832	.0871	.0910	.0948	.0987	.1026	.1064	.1103	.1141
.3	.1179	.1217	.1255	.1293	.1331	.1368	.1406	.1443	.1480	.1517
.4	.1554	.1591	.1628	.1664	.1700	.1736	.1772	.1808	.1844	.1879
.5	.1915	.1950	.1985	.2019	.2054	.2088	.2123	.2157	.2190	.2224
.6	.2257	.2291	.2324	.2357	.2389	.2422	.2454	.2486	.2517	.2549
.7	.2580	.2611	.2642	.2673	.2704	.2734	.2764	.2794	.2823	.2852
.8	.2881	.2910	.2939	.2967	.2995	.3023	.3051	.3078	.3106	.3133
.9	.3159	.3186	.3212	.3238	.3264	.3289	.3315	.3340	.3365	.3389
1.0	.3413	.3438	.3461	.3485	.3508	.3531	.3554	.3577	.3599	.3621
1.1	.3643	.3665	.3686	.3708	.3729	.3749	.3770	.3790	.3810	.3830
1.2	.3849	.3869	.3888	.3907	.3925	.3944	.3962	.3980	.3997	.4015
1.3	.4032	.4049	.4066	.4082	.4099	.4115	.4131	.4147	.4162	.4177
1.4	.4192	.4207	.4222	.4236	.4251	.4265	.4279	.4292	.4306	.4319
1.5	.4332	.4345	.4357	.4370	.4382	.4394	.4406	.4418	.4429	.4441
1.6	.4452	.4463	.4474	.4484	.4495	.4505	.4515	.4525	.4535	.4545
1.7	.4554	.4564	.4573	.4582	.4591	.4599	.4608	.4616	.4625	.4633
1.8	.4641	.4649	.4656	.4664	.4671	.4678	.4686	.4693	.4699	.4706
1.9	.4713	.4719	.4726	.4732	.4738	.4744	.4750	.4756	.4761	.4767
2.0	.4772	.4778	.4783	.4788	.4793	.4798	.4803	.4808	.4812	.4817
2.1	.4821	.4826	.4830	.4834	.4838	.4842	.4846	.4850	.4854	.4857
2.2	.4861	.4864	.4868	.4871	.4875	.4878	.4881	.4884	.4887	.4890
2.3	.4893	.4896	.4898	.4901	.4904	.4906	.4909	.4911	.4913	.4916
2.4	.4918	.4920	.4922	.4925	.4927	.4929	.4931	.4932	.4934	.4936
2.5	.4938	.4940	.4941	.4943	.4945	.4946	.4948	.4949	.4951	.4952
2.6	.4953	.4955	.4956	.4957	.4959	.4960	.4961	.4962	.4963	.4964
2.7	.4965	.4966	.4967	.4968	.4969	.4970	.4971	.4972	.4973	.4974
2.8	.4974	.4975	.4976	.4977	.4977	.4978	.4979	.4979	.4980	.4981
2.9	.4981	.4982	.4982	.4983	.4984	.4984	.4985	.4985	.4986	.4986
3.0	.4987	.4987	.4987	.4988	.4988	.4989	.4989	.4989	.4990	.4990
3.1	.4990	.4991	.4991	.4991	.4992	.4992	.4992	.4992	.4993	.4993
3.2	.4993	.4993	.4994	.4994	.4994	.4994	.4994	.4995	.4995	.4995
3.3	.4995	.4995	.4995	.4996	.4996	.4996	.4996	.4996	.4996	.4997
3.4	.4997	.4997	.4997	.4997	.4997	.4997	.4997	.4997	.4997	.4998
3.5	.4998									
4.0	.49997									
4.5	.499997									
5.0	.4999997									

Reprinted from John T. Roscoe, *Fundamental Research Statistics for the Behavioral Sciences,* Second Edition (New York: Holt, Rinehart and Winston, 1975, p. 425) by permission of the author and publisher.

Table C. Critical values of the *t* and *z* distributions

Statistic	Degrees of freedom	.05	.025	.01	.005	.0005	Alpha level for one-tail test†
		.10	.05	.02	.01	.001	Alpha level for two-tail test
t	1	6.314	12.706	31.821	63.657	636.619	
	2	2.920	4.303	6.965	9.925	31.598	
	3	2.353	3.182	4.541	5.841	12.941	
	4	2.132	2.776	3.747	4.604	8.610	
	5	2.015	2.571	3.365	4.032	6.859	
	6	1.943	2.447	3.143	3.707	5.959	
	7	1.895	2.365	2.998	3.499	5.405	
	8	1.860	2.306	2.896	3.355	5.041	
	9	1.833	2.262	2.821	3.250	4.781	
	10	1.812	2.228	2.764	3.169	4.587	
	11	1.796	2.201	2.718	3.106	4.437	
	12	1.782	2.179	2.681	3.055	4.318	
	13	1.771	2.160	2.650	3.012	4.221	
	14	1.761	2.145	2.624	2.977	4.140	
	15	1.753	2.131	2.602	2.947	4.073	
	16	1.746	2.120	2.583	2.921	4.015	
	17	1.740	2.110	2.567	2.898	3.965	
	18	1.734	2.101	2.552	2.878	3.922	
	19	1.729	2.093	2.539	2.861	3.883	
	20	1.725	2.086	2.528	2.845	3.850	
	21	1.721	2.080	2.518	2.831	3.819	
	22	1.717	2.074	2.508	2.819	3.792	
	23	1.714	2.069	2.500	2.807	3.767	
	24	1.711	2.064	2.492	2.797	3.745	
	25	1.708	2.060	2.485	2.787	3.725	
	26	1.706	2.056	2.479	2.779	3.707	
	27	1.703	2.052	2.473	2.771	3.690	
	28	1.701	2.048	2.467	2.763	3.674	
	29	1.699	2.045	2.462	2.756	3.659	
	30	1.697	2.042	2.457	2.750	3.646	
	40	1.684	2.021	2.423	2.704	3.551	
	60	1.671	2.000	2.310	2.660	3.460	
	120	1.658	1.980	2.358	2.617	3.373	
z		1.645	1.960	2.326	2.576	3.291	

†Note: for a one-tail test, tabulated values are to be preceded by either a + or a − sign. For a two-tail test, precede tabulated values by a ± to indicate sample values at either end of the distribution can be used to reject the null hypothesis.

Table D. Critical values of the χ^2 and z distributions

df	0.995	0.990	0.975	0.950	0.900	0.750	0.500
1	392704.10^{-10}	157088.10^{-9}	982069.10^{-9}	393214.10^{-8}	0.0157908	0.1015308	0.454937
2	0.0100251	0.0201007	0.0506356	0.102587	0.210720	0.575364	1.38629
3	0.0717212	0.114832	0.215795	0.351846	0.584375	1.212534	2.36597
4	0.206990	0.297110	0.484419	0.710721	1.063623	1.92255	3.35670
5	0.411740	0.554300	0.831211	1.145476	1.61031	2.67460	4.35146
6	0.675727	0.872085	1.237347	1.63539	2.20413	3.45460	5.34812
7	0.989265	1.239043	1.68987	2.16735	2.83311	4.25485	6.34581
8	1.344419	1.646482	2.17973	2.73264	3.48954	5.07064	7.34412
9	1.734926	2.087912	2.70039	3.32511	4.16816	5.89883	8.34283
10	2.15585	2.55821	3.24697	3.94030	4.86518	6.73720	9.34182
11	2.60321	3.05347	3.81575	4.57481	5.57779	7.58412	10.3410
12	3.07382	3.57056	4.40379	5.22603	6.30380	8.43842	11.3403
13	3.56503	4.10691	5.00874	5.89186	7.04150	9.29906	12.3398
14	4.07468	4.66043	5.62872	6.57063	7.78953	10.1653	13.3393
15	4.60094	5.22935	6.26214	7.26094	8.54675	11.0365	14.3389
16	5.14224	5.81221	6.90766	7.96164	9.31223	11.9122	15.3385
17	5.69724	6.40776	7.56418	8.67176	10.0852	12.7919	16.3381
18	6.26481	7.01491	8.23075	9.39046	10.8649	13.6753	17.3379
19	6.84398	7.63273	8.90655	10.1170	11.6509	14.5620	18.3376
20	7.43386	8.26040	9.59083	10.8508	12.4426	15.4518	19.3374
21	8.03366	8.89720	10.28293	11.5913	13.2396	16.3444	20.3372
22	8.64272	9.54249	10.9823	12.3380	14.0415	17.2396	21.3370
23	9.26042	10.19567	11.6885	13.0905	14.8479	18.1373	22.3369
24	9.88623	10.8564	12.4011	13.8484	15.6587	19.0372	23.3367
25	10.5197	11.5240	13.1197	14.6114	16.4734	19.9393	24.3366
26	11.1603	12.1981	13.8439	15.3791	17.2919	20.8434	25.3364
27	11.8076	12.8786	14.5733	16.1513	18.1138	21.7494	26.3363
28	12.4613	13.5648	15.3079	16.9279	18.9392	22.6572	27.3363
29	13.1211	14.2565	16.0471	17.7083	19.7677	23.5666	28.3362
30	13.7867	14.9535	16.7908	18.4926	20.5992	24.4776	29.3360
40	20.7065	22.1643	24.4331	26.5093	29.0505	33.6603	39.3354
50	27.9907	29.7067	32.3574	34.7642	37.6886	42.9421	49.3349
60	35.5346	37.4848	40.4817	43.1879	46.4589	52.2938	59.3347
70	43.2752	45.4418	48.7576	51.7393	55.3290	61.6983	69.3344
80	51.1720	53.5400	57.1532	60.3915	64.2778	71.1445	79.3343
90	59.1963	61.7541	65.6466	69.1260	73.2912	80.6247	89.3342
100	67.3276	70.0648	74.2219	77.9295	82.3581	90.1332	99.3341
z	-2.5758	-2.3263	-1.9600	-1.6449	-1.2816	-0.6745	0.0000

df	0.250	0.100	0.050	0.025	0.010	0.005	0.001
1	1.32330	2.70554	3.84146	5.02389	6.63490	7.87944	10.828
2	2.77259	4.60517	5.99147	7.37776	9.21034	10.5966	13.816
3	4.10835	6.25139	7.81473	9.34840	11.3449	12.8381	16.266
4	5.38527	7.77944	9.48773	11.1433	13.2767	14.8602	18.467
5	6.62568	9.23635	11.0705	12.8325	15.0863	16.7496	20.515
6	7.84080	10.6446	12.5916	14.4494	16.8119	18.5476	22.458
7	9.03715	12.0170	14.0671	16.0128	18.4753	20.2777	24.322
8	10.2188	13.3616	15.5073	17.5346	20.0902	21.9550	26.125
9	11.3887	14.6837	16.9190	19.0228	21.6660	23.5893	27.877
10	12.5489	15.9871	18.3070	20.4831	23.2093	25.1882	29.588
11	13.7007	17.2750	19.6751	21.9200	24.7250	26.7569	31.264
12	14.8454	18.5494	21.0261	23.3367	26.2170	28.2995	32.909
13	15.9839	19.8119	22.3621	24.7356	27.6883	29.8194	34.528
14	17.1170	21.0642	23.6848	26.1190	29.1413	31.3193	36.123
15	18.2451	22.3072	24.9958	27.4884	30.5779	32.8013	37.697
16	19.3688	23.5418	26.2962	28.8454	31.9999	34.2672	39.252
17	20.4887	24.7690	27.5871	30.1910	33.4087	35.7185	40.790
18	21.6049	25.9894	28.8693	31.5264	34.8053	37.1564	42.312
19	22.7178	27.2036	30.1435	32.8523	36.1908	38.5822	43.820
20	23.8277	28.4120	31.4104	34.1696	37.5662	39.9968	45.315
21	24.9348	29.6151	32.6705	35.4789	38.9321	41.4010	46.797
22	26.0393	30.8133	33.9244	36.7807	40.2894	42.7956	48.268
23	27.1413	32.0069	35.1725	38.0757	41.6384	44.1813	49.728
24	28.2412	33.1963	36.4151	39.3641	42.9798	45.5585	51.179
25	29.3389	34.3816	37.6525	40.6465	44.3141	46.9278	52.620
26	30.4345	35.5631	38.8852	41.9232	45.6417	48.2899	54.052
27	31.5284	36.7412	40.1133	43.1944	46.9630	49.6449	55.476
28	32.6205	37.9159	41.3372	44.4607	48.2782	50.9933	56.892
29	33.7109	39.0875	42.5569	45.7222	49.5879	52.3356	58.302
30	34.7998	40.2560	43.7729	46.9792	50.8922	53.6720	59.703
40	45.6160	51.8050	55.7585	59.3417	63.6907	66.7659	73.402
50	56.3336	63.1671	67.5048	71.4202	76.1539	79.4900	86.661
60	66.9814	74.3970	79.0819	83.2976	88.3794	91.9517	99.607
70	77.5766	85.5271	90.5312	95.0231	100.425	104.215	112.317
80	88.1303	96.5782	101.879	106.629	112.329	116.321	124.839
90	98.6499	107.565	113.145	118.136	124.116	128.299	137.208
100	109.141	118.498	124.342	129.561	135.807	140.169	149.449
z	+0.6745	+1.2816	+1.6449	+1.9600	+2.3263	+2.5758	+3.0902

Table E. Transformation of r to Z

r	z_r	r	z_r	r	z_r	r	z_r	r	z_r
.000	.000	.200	.203	.400	.424	.600	.693	.800	1.099
.005	.005	.205	.208	.405	.430	.605	.701	.805	1.113
.010	.010	.210	.213	.410	.436	.610	.709	.810	1.127
.015	.015	.215	.218	.415	.442	.615	.717	.815	1.142
.020	.020	.220	.224	.420	.448	.620	.725	.820	1.157
.025	.025	.225	.229	.425	.454	.625	.733	.825	1.172
.030	.030	.230	.234	.430	.460	.630	.741	.830	1.188
.035	.035	.235	.239	.435	.466	.635	.750	.835	1.204
.040	.040	.240	.245	.440	.472	.640	.758	.840	1.221
.045	.045	.245	.250	.445	.478	.645	.767	.845	1.238
.050	.050	.250	.255	.450	.485	.650	.775	.850	1.256
.055	.055	.255	.261	.455	.491	.655	.784	.855	1.274
.060	.060	.260	.266	.460	.497	.660	.793	.860	1.293
.065	.065	.265	.271	.465	.504	.665	.802	.865	1.313
.070	.070	.270	.277	.470	.510	.670	.811	.870	1.333
.075	.075	.275	.282	.475	.517	.675	.820	.875	1.354
.080	.080	.280	.288	.480	.523	.680	.829	.880	1.376
.085	.085	.285	.293	.485	.530	.685	.838	.885	1.398
.090	.090	.290	.299	.490	.536	.690	.848	.890	1.422
.095	.095	.295	.304	.495	.543	.695	.858	.895	1.447
.100	.100	.300	.310	.500	.549	.700	.867	.900	1.472
.105	.105	.305	.315	.505	.556	.705	.877	.905	1.499
.110	.110	.310	.321	.510	.563	.710	.887	.910	1.528
.115	.116	.315	.326	.515	.570	.715	.897	.915	1.557
.120	.121	.320	.332	.520	.576	.720	.908	.920	1.589
.125	.126	.325	.337	.525	.583	.725	.918	.925	1.623
.130	.131	.330	.343	.530	.590	.730	.929	.930	1.658
.135	.136	.335	.348	.535	.597	.735	.940	.935	1.697
.140	.141	.340	.354	.540	.604	.740	.950	.940	1.738
.145	.146	.345	.360	.545	.611	.745	.962	.945	1.783
.150	.151	.350	.365	.550	.618	.750	.973	.950	1.832
.155	.156	.355	.371	.555	.626	.755	.984	.955	1.886
.160	.161	.360	.377	.560	.633	.760	.996	.960	1.946
.165	.167	.365	.383	.565	.640	.765	1.008	.965	2.014
.170	.172	.370	.388	.570	.648	.770	1.020	.970	2.092
.175	.177	.375	.394	.575	.655	.775	1.033	.975	2.185
.180	.182	.380	.400	.580	.662	.780	1.045	.980	2.298
.185	.187	.385	.406	.585	.670	.785	1.058	.985	2.443
.190	.192	.390	.412	.590	.678	.790	1.071	.990	2.647
.195	.198	.395	.418	.595	.685	.795	1.085	.995	2.994

The obtained F is significant at a given level if it is equal to or greater than the value shown in the table.
0.05 (light row) and 0.01 (dark row) points for the distribution of F

df (lesser)	\ df (greater, numerator) →	1	2	3	4	5	6	7	8	9	10	11	12	14	16	20	24	30	40	50	75	100	200	500	∞
1		161 / 4052	200 / 4999	216 / 5403	225 / 5625	230 / 5764	234 / 5859	237 / 5928	239 / 5981	241 / 6022	242 / 6056	243 / 6082	244 / 6106	245 / 6142	246 / 6169	248 / 6208	249 / 6234	250 / 6258	251 / 6286	252 / 6302	253 / 6323	253 / 6334	254 / 6352	254 / 6361	254 / 6366
2		18.51 / 98.49	19.00 / 99.01	19.16 / 99.17	19.25 / 99.25	19.30 / 99.30	19.33 / 99.33	19.36 / 99.34	19.37 / 99.36	19.38 / 99.38	19.39 / 99.40	19.40 / 99.41	19.41 / 99.42	19.42 / 99.43	19.43 / 99.44	19.44 / 99.45	19.45 / 99.46	19.46 / 99.47	19.47 / 99.48	19.47 / 99.48	19.48 / 99.49	19.49 / 99.49	19.49 / 99.49	19.50 / 99.50	19.50 / 99.50
3		10.13 / 34.12	9.55 / 30.81	9.28 / 29.46	9.12 / 28.71	9.01 / 28.24	8.94 / 27.91	8.88 / 27.67	8.84 / 27.49	8.81 / 27.34	8.78 / 27.23	8.76 / 27.13	8.74 / 27.05	8.71 / 26.92	8.69 / 26.83	8.66 / 26.69	8.64 / 26.60	8.62 / 26.50	8.60 / 26.41	8.58 / 26.30	8.57 / 26.27	8.56 / 26.23	8.54 / 26.18	8.54 / 26.14	8.53 / 26.12
4		7.71 / 21.20	6.94 / 18.00	6.59 / 16.69	6.39 / 15.98	6.26 / 15.52	6.16 / 15.21	6.09 / 14.98	6.04 / 14.80	6.00 / 14.66	5.96 / 14.54	5.93 / 14.45	5.91 / 14.37	5.87 / 14.24	5.84 / 14.15	5.80 / 14.02	5.77 / 13.93	5.74 / 13.83	5.71 / 13.74	5.70 / 13.69	5.68 / 13.61	5.66 / 13.57	5.65 / 13.52	5.64 / 13.48	5.63 / 13.46
5		6.61 / 16.26	5.79 / 13.27	5.41 / 12.06	5.19 / 11.39	5.05 / 10.97	4.95 / 10.67	4.88 / 10.45	4.82 / 10.27	4.78 / 10.15	4.74 / 10.05	4.70 / 9.96	4.68 / 9.89	4.64 / 9.77	4.60 / 9.68	4.56 / 9.55	4.53 / 9.47	4.50 / 9.38	4.46 / 9.29	4.44 / 9.24	4.42 / 9.17	4.40 / 9.13	4.38 / 9.07	4.37 / 9.04	4.36 / 9.02
6		5.99 / 13.74	5.14 / 10.92	4.76 / 9.78	4.53 / 9.15	4.39 / 8.75	4.28 / 8.47	4.21 / 8.26	4.15 / 8.10	4.10 / 7.98	4.06 / 7.87	4.03 / 7.79	4.00 / 7.72	3.96 / 7.60	3.92 / 7.52	3.87 / 7.39	3.84 / 7.31	3.81 / 7.23	3.77 / 7.14	3.75 / 7.09	3.72 / 7.02	3.71 / 6.99	3.69 / 6.94	3.68 / 6.90	3.67 / 6.88
7		5.59 / 12.25	4.74 / 9.55	4.35 / 8.45	4.12 / 7.85	3.97 / 7.46	3.87 / 7.19	3.79 / 7.00	3.73 / 6.84	3.68 / 6.71	3.63 / 6.62	3.60 / 6.54	3.57 / 6.47	3.52 / 6.35	3.49 / 6.27	3.44 / 6.15	3.41 / 6.07	3.38 / 5.98	3.34 / 5.90	3.32 / 5.85	3.29 / 5.78	3.28 / 5.75	3.25 / 5.70	3.24 / 5.67	3.23 / 5.65
8		5.32 / 11.26	4.46 / 8.65	4.07 / 7.59	3.84 / 7.01	3.69 / 6.63	3.58 / 6.37	3.50 / 6.19	3.44 / 6.03	3.39 / 5.91	3.34 / 5.82	3.31 / 5.74	3.28 / 5.67	3.23 / 5.56	3.20 / 5.48	3.15 / 5.36	3.12 / 5.28	3.08 / 5.20	3.05 / 5.11	3.03 / 5.06	3.00 / 5.00	2.98 / 4.96	2.96 / 4.91	2.94 / 4.88	2.93 / 4.86
9		5.12 / 10.56	4.26 / 8.02	3.86 / 6.99	3.63 / 6.42	3.48 / 6.06	3.37 / 5.80	3.29 / 5.62	3.23 / 5.47	3.18 / 5.35	3.13 / 5.26	3.10 / 5.18	3.07 / 5.11	3.02 / 5.00	2.98 / 4.92	2.93 / 4.80	2.90 / 4.73	2.86 / 4.64	2.82 / 4.56	2.80 / 4.51	2.77 / 4.45	2.76 / 4.41	2.73 / 4.36	2.72 / 4.33	2.71 / 4.31
10		4.96 / 10.04	4.10 / 7.56	3.71 / 6.55	3.48 / 5.99	3.33 / 5.64	3.22 / 5.39	3.14 / 5.21	3.07 / 5.06	3.02 / 4.95	2.97 / 4.85	2.94 / 4.78	2.91 / 4.71	2.86 / 4.60	2.82 / 4.52	2.77 / 4.41	2.74 / 4.33	2.70 / 4.25	2.67 / 4.17	2.64 / 4.12	2.61 / 4.05	2.59 / 4.01	2.56 / 3.96	2.55 / 3.93	2.54 / 3.91
11		4.84 / 9.65	3.98 / 7.20	3.59 / 6.22	3.36 / 5.67	3.20 / 5.32	3.09 / 5.07	3.01 / 4.88	2.95 / 4.74	2.90 / 4.63	2.86 / 4.54	2.82 / 4.46	2.79 / 4.40	2.74 / 4.29	2.70 / 4.21	2.65 / 4.10	2.61 / 4.02	2.57 / 3.94	2.53 / 3.86	2.50 / 3.80	2.47 / 3.74	2.45 / 3.70	2.42 / 3.66	2.41 / 3.62	2.40 / 3.60
12		4.75 / 9.33	3.88 / 6.93	3.49 / 5.95	3.26 / 5.41	3.11 / 5.06	3.00 / 4.82	2.92 / 4.65	2.85 / 4.50	2.80 / 4.39	2.76 / 4.30	2.72 / 4.22	2.69 / 4.16	2.64 / 4.05	2.60 / 3.98	2.54 / 3.86	2.50 / 3.78	2.46 / 3.70	2.42 / 3.61	2.40 / 3.56	2.36 / 3.49	2.35 / 3.46	2.32 / 3.41	2.31 / 3.38	2.30 / 3.36
13		4.67 / 9.07	3.80 / 6.70	3.41 / 5.74	3.18 / 5.20	3.02 / 4.86	2.92 / 4.62	2.84 / 4.44	2.77 / 4.30	2.72 / 4.19	2.67 / 4.10	2.63 / 4.02	2.60 / 3.96	2.55 / 3.85	2.51 / 3.78	2.46 / 3.67	2.42 / 3.59	2.38 / 3.51	2.34 / 3.42	2.32 / 3.37	2.28 / 3.30	2.26 / 3.27	2.24 / 3.21	2.22 / 3.18	2.21 / 3.16
14		4.60 / 8.86	3.74 / 6.51	3.34 / 5.56	3.11 / 5.03	2.96 / 4.69	2.85 / 4.46	2.77 / 4.28	2.70 / 4.14	2.65 / 4.03	2.60 / 3.94	2.56 / 3.86	2.53 / 3.80	2.48 / 3.70	2.44 / 3.62	2.39 / 3.51	2.35 / 3.43	2.31 / 3.34	2.27 / 3.26	2.24 / 3.21	2.21 / 3.14	2.19 / 3.11	2.16 / 3.06	2.14 / 3.02	2.13 / 3.00
15		4.54 / 8.68	3.68 / 6.36	3.29 / 5.42	3.06 / 4.89	2.90 / 4.56	2.79 / 4.32	2.70 / 4.14	2.64 / 4.00	2.59 / 3.89	2.55 / 3.80	2.51 / 3.73	2.48 / 3.67	2.43 / 3.56	2.39 / 3.48	2.33 / 3.36	2.29 / 3.29	2.25 / 3.20	2.21 / 3.12	2.18 / 3.07	2.15 / 3.00	2.12 / 2.97	2.10 / 2.92	2.08 / 2.89	2.07 / 2.87

Degrees of freedom for lesser mean square

(continued)

The obtained F is significant at a given level if it is equal to or greater than the value shown in the table.
0.05 (light row) and 0.01 (dark row) points for the distribution of F

Degrees of freedom for greater mean square (numerator). Cells show light row (0.05) / dark row (0.01).

df (lesser)	∞	500	200	100	75	50	40	30	24	20	16	14	12	11	10	9	8	7	6	5	4	3	2	1
16	2.01 / 2.75	2.02 / 2.77	2.04 / 2.80	2.07 / 2.86	2.09 / 2.89	2.13 / 2.96	2.16 / 3.01	2.20 / 3.10	2.24 / 3.18	2.28 / 3.25	2.33 / 3.37	2.37 / 3.45	2.42 / 3.55	2.45 / 3.61	2.49 / 3.69	2.54 / 3.78	2.59 / 3.89	2.66 / 4.03	2.74 / 4.20	2.85 / 4.44	3.01 / 4.77	3.24 / 5.29	3.63 / 6.23	4.49 / 8.53
17	1.96 / 2.65	1.97 / 2.67	1.99 / 2.70	2.02 / 2.76	2.04 / 2.79	2.08 / 2.86	2.11 / 2.92	2.15 / 3.00	2.19 / 3.08	2.23 / 3.16	2.29 / 3.27	2.33 / 3.35	2.38 / 3.45	2.41 / 3.52	2.45 / 3.59	2.50 / 3.68	2.55 / 3.79	2.62 / 3.93	2.70 / 4.10	2.81 / 4.34	2.96 / 4.67	3.20 / 5.18	3.59 / 6.11	4.45 / 8.40
18	1.92 / 2.57	1.93 / 2.59	1.95 / 2.62	1.98 / 2.68	2.00 / 2.71	2.04 / 2.78	2.07 / 2.83	2.11 / 2.91	2.15 / 3.00	2.19 / 3.07	2.25 / 3.19	2.29 / 3.27	2.34 / 3.37	2.37 / 3.44	2.41 / 3.51	2.46 / 3.60	2.51 / 3.71	2.58 / 3.85	2.66 / 4.01	2.77 / 4.25	2.93 / 4.58	3.16 / 5.09	3.55 / 6.01	4.41 / 8.28
19	1.88 / 2.49	1.90 / 2.51	1.91 / 2.54	1.94 / 2.60	1.96 / 2.63	2.00 / 2.70	2.02 / 2.76	2.07 / 2.84	2.11 / 2.92	2.15 / 3.00	2.21 / 3.12	2.26 / 3.19	2.31 / 3.30	2.34 / 3.36	2.38 / 3.43	2.43 / 3.52	2.48 / 3.63	2.55 / 3.77	2.63 / 3.94	2.74 / 4.17	2.90 / 4.50	3.13 / 5.01	3.52 / 5.93	4.38 / 8.18
20	1.84 / 2.42	1.85 / 2.44	1.87 / 2.47	1.90 / 2.53	1.92 / 2.56	1.96 / 2.63	1.99 / 2.69	2.04 / 2.77	2.08 / 2.86	2.12 / 2.94	2.18 / 3.05	2.23 / 3.13	2.28 / 3.23	2.31 / 3.30	2.35 / 3.37	2.40 / 3.45	2.45 / 3.56	2.52 / 3.71	2.60 / 3.87	2.71 / 4.10	2.87 / 4.43	3.10 / 4.94	3.49 / 5.85	4.35 / 8.10
21	1.81 / 2.36	1.82 / 2.38	1.84 / 2.42	1.87 / 2.47	1.90 / 2.51	1.93 / 2.58	1.96 / 2.63	2.00 / 2.72	2.05 / 2.80	2.09 / 2.88	2.15 / 2.99	2.20 / 3.07	2.25 / 3.17	2.28 / 3.24	2.32 / 3.31	2.37 / 3.40	2.42 / 3.51	2.49 / 3.65	2.57 / 3.81	2.68 / 4.04	2.84 / 4.37	3.07 / 4.87	3.47 / 5.78	4.32 / 8.02
22	1.78 / 2.31	1.80 / 2.33	1.81 / 2.37	1.84 / 2.42	1.87 / 2.46	1.91 / 2.53	1.93 / 2.58	1.98 / 2.67	2.03 / 2.75	2.07 / 2.83	2.13 / 2.94	2.18 / 3.02	2.23 / 3.12	2.26 / 3.18	2.30 / 3.26	2.35 / 3.35	2.40 / 3.45	2.47 / 3.59	2.55 / 3.76	2.66 / 3.99	2.82 / 4.31	3.05 / 4.82	3.44 / 5.72	4.30 / 7.94
23	1.76 / 2.26	1.77 / 2.28	1.79 / 2.32	1.82 / 2.37	1.84 / 2.41	1.88 / 2.48	1.91 / 2.53	1.96 / 2.62	2.00 / 2.70	2.04 / 2.78	2.10 / 2.89	2.14 / 2.97	2.20 / 3.07	2.24 / 3.14	2.28 / 3.21	2.32 / 3.30	2.38 / 3.41	2.45 / 3.54	2.53 / 3.71	2.64 / 3.94	2.80 / 4.26	3.03 / 4.76	3.42 / 5.66	4.28 / 7.88
24	1.73 / 2.21	1.74 / 2.23	1.76 / 2.27	1.80 / 2.33	1.82 / 2.36	1.86 / 2.44	1.89 / 2.49	1.94 / 2.58	1.98 / 2.66	2.02 / 2.74	2.09 / 2.85	2.13 / 2.93	2.18 / 3.03	2.22 / 3.09	2.26 / 3.17	2.30 / 3.25	2.36 / 3.36	2.43 / 3.50	2.51 / 3.67	2.62 / 3.90	2.78 / 4.22	3.01 / 4.72	3.40 / 5.61	4.26 / 7.82
25	1.71 / 2.17	1.72 / 2.19	1.74 / 2.23	1.77 / 2.29	1.80 / 2.32	1.84 / 2.40	1.87 / 2.45	1.92 / 2.54	1.96 / 2.62	2.00 / 2.70	2.06 / 2.81	2.11 / 2.89	2.16 / 2.99	2.20 / 3.05	2.24 / 3.13	2.28 / 3.21	2.34 / 3.32	2.41 / 3.46	2.49 / 3.63	2.60 / 3.86	2.76 / 4.18	2.99 / 4.68	3.38 / 5.57	4.24 / 7.77
26	1.69 / 2.13	1.70 / 2.15	1.72 / 2.19	1.76 / 2.25	1.78 / 2.28	1.82 / 2.36	1.85 / 2.41	1.90 / 2.50	1.95 / 2.58	1.99 / 2.66	2.05 / 2.77	2.10 / 2.86	2.15 / 2.96	2.18 / 3.02	2.22 / 3.09	2.27 / 3.17	2.32 / 3.29	2.39 / 3.42	2.47 / 3.59	2.59 / 3.82	2.74 / 4.14	2.98 / 4.64	3.37 / 5.53	4.22 / 7.72
27	1.67 / 2.10	1.68 / 2.12	1.71 / 2.16	1.74 / 2.21	1.76 / 2.25	1.80 / 2.33	1.84 / 2.38	1.88 / 2.47	1.93 / 2.55	1.97 / 2.63	2.03 / 2.74	2.08 / 2.83	2.13 / 2.93	2.16 / 2.98	2.20 / 3.06	2.25 / 3.14	2.30 / 3.26	2.37 / 3.39	2.46 / 3.56	2.57 / 3.79	2.73 / 4.11	2.96 / 4.60	3.35 / 5.49	4.21 / 7.68
28	1.65 / 2.06	1.67 / 2.09	1.69 / 2.13	1.72 / 2.18	1.75 / 2.22	1.78 / 2.30	1.81 / 2.35	1.87 / 2.44	1.91 / 2.52	1.96 / 2.60	2.02 / 2.71	2.06 / 2.80	2.12 / 2.90	2.15 / 2.95	2.19 / 3.03	2.24 / 3.11	2.29 / 3.23	2.36 / 3.36	2.44 / 3.53	2.56 / 3.76	2.71 / 4.07	2.95 / 4.57	3.34 / 5.45	4.20 / 7.64
29	1.64 / 2.03	1.65 / 2.06	1.68 / 2.10	1.71 / 2.15	1.73 / 2.19	1.77 / 2.27	1.80 / 2.32	1.85 / 2.41	1.90 / 2.49	1.94 / 2.57	2.00 / 2.68	2.05 / 2.77	2.10 / 2.87	2.14 / 2.92	2.18 / 3.00	2.22 / 3.08	2.28 / 3.20	2.35 / 3.32	2.43 / 3.50	2.54 / 3.73	2.70 / 4.04	2.93 / 4.54	3.33 / 5.42	4.18 / 7.60
30	1.62 / 2.01	1.64 / 2.03	1.66 / 2.07	1.69 / 2.13	1.72 / 2.16	1.76 / 2.24	1.79 / 2.29	1.84 / 2.38	1.89 / 2.47	1.93 / 2.55	1.99 / 2.66	2.04 / 2.74	2.09 / 2.84	2.12 / 2.90	2.16 / 2.98	2.21 / 3.06	2.27 / 3.17	2.34 / 3.30	2.42 / 3.47	2.53 / 3.70	2.69 / 4.02	2.92 / 4.51	3.32 / 5.39	4.17 / 7.56

Degrees of freedom for lesser mean square

0.05 (light row) and 0.01 (dark row) points for the distribution of F

Degrees of freedom for greater mean square (numerator)

Degrees of freedom for lesser mean square

df	1	2	3	4	5	6	7	8	9	10	11	12	14	16	20	24	30	40	50	75	100	200	500	∞
32	4.15	3.30	2.90	2.67	2.51	2.40	2.32	2.25	2.19	2.14	2.10	2.07	2.02	1.97	1.91	1.86	1.82	1.76	1.74	1.69	1.67	1.64	1.61	1.59
	7.50	5.34	4.46	3.97	3.66	3.42	3.25	3.12	3.01	2.94	2.86	2.80	2.70	2.62	2.51	2.42	2.34	2.25	2.20	2.12	2.08	2.02	1.98	1.96
34	4.13	3.28	2.88	2.65	2.49	2.38	2.30	2.23	2.17	2.12	2.08	2.05	2.00	1.95	1.89	1.84	1.80	1.74	1.71	1.67	1.64	1.61	1.59	1.57
	7.44	5.29	4.42	3.93	3.61	3.38	3.21	3.08	2.97	2.89	2.82	2.76	2.66	2.58	2.47	2.38	2.30	2.21	2.15	2.08	2.04	1.98	1.94	1.91
36	4.11	3.26	2.86	2.63	2.48	2.36	2.28	2.21	2.15	2.10	2.06	2.03	1.98	1.93	1.87	1.82	1.78	1.72	1.69	1.65	1.62	1.59	1.56	1.55
	7.39	5.25	4.38	3.89	3.58	3.35	3.18	3.04	2.94	2.86	2.78	2.72	2.62	2.54	2.43	2.35	2.26	2.17	2.12	2.04	2.00	1.94	1.90	1.87
38	4.10	3.25	2.85	2.62	2.46	2.35	2.26	2.19	2.14	2.09	2.05	2.02	1.96	1.92	1.85	1.80	1.76	1.71	1.67	1.63	1.60	1.57	1.54	1.53
	7.35	5.21	4.34	3.86	3.54	3.32	3.15	3.02	2.91	2.82	2.75	2.69	2.59	2.51	2.40	2.32	2.22	2.14	2.08	2.00	1.97	1.90	1.86	1.84
40	4.08	3.23	2.84	2.61	2.45	2.34	2.25	2.18	2.12	2.07	2.04	2.00	1.95	1.90	1.84	1.79	1.74	1.69	1.66	1.61	1.59	1.55	1.53	1.51
	7.31	5.18	4.31	3.83	3.51	3.29	3.12	2.99	2.88	2.80	2.73	2.66	2.56	2.49	2.37	2.29	2.20	2.11	2.05	1.97	1.94	1.88	1.84	1.81
42	4.07	3.22	2.83	2.59	2.44	2.32	2.24	2.17	2.11	2.06	2.02	1.99	1.94	1.89	1.82	1.78	1.73	1.68	1.64	1.60	1.57	1.54	1.51	1.49
	7.27	5.15	4.29	3.80	3.49	3.26	3.10	2.96	2.86	2.77	2.70	2.64	2.54	2.46	2.35	2.26	2.17	2.08	2.02	1.94	1.91	1.85	1.80	1.78
44	4.06	3.21	2.82	2.58	2.43	2.31	2.23	2.16	2.10	2.05	2.01	1.98	1.92	1.88	1.81	1.76	1.72	1.66	1.63	1.58	1.56	1.52	1.50	1.48
	7.24	5.12	4.26	3.78	3.46	3.24	3.07	2.94	2.84	2.75	2.68	2.62	2.52	2.44	2.32	2.24	2.15	2.06	2.00	1.92	1.88	1.82	1.78	1.75
46	4.05	3.20	2.81	2.57	2.42	2.30	2.22	2.14	2.09	2.04	2.00	1.97	1.91	1.87	1.80	1.75	1.71	1.65	1.62	1.57	1.54	1.51	1.48	1.46
	7.21	5.10	4.24	3.76	3.44	3.22	3.05	2.92	2.82	2.73	2.66	2.60	2.50	2.42	2.30	2.22	2.13	2.04	1.98	1.90	1.86	1.80	1.76	1.72
48	4.04	3.19	2.80	2.56	2.41	2.30	2.21	2.14	2.08	2.03	1.99	1.96	1.90	1.86	1.79	1.74	1.70	1.64	1.61	1.56	1.53	1.50	1.47	1.45
	7.19	5.08	4.22	3.74	3.42	3.20	3.04	2.90	2.80	2.71	2.64	2.58	2.48	2.40	2.28	2.20	2.11	2.02	1.96	1.88	1.84	1.78	1.73	1.70
50	4.03	3.18	2.79	2.56	2.40	2.29	2.20	2.13	2.07	2.02	1.98	1.95	1.90	1.85	1.78	1.74	1.69	1.63	1.60	1.55	1.52	1.48	1.46	1.44
	7.17	5.06	4.20	3.72	3.41	3.18	3.02	2.88	2.78	2.70	2.62	2.56	2.46	2.39	2.26	2.18	2.10	2.00	1.94	1.86	1.82	1.76	1.71	1.68
55	4.02	3.17	2.78	2.54	2.38	2.27	2.18	2.11	2.05	2.00	1.97	1.93	1.88	1.83	1.76	1.72	1.67	1.61	1.58	1.52	1.50	1.46	1.43	1.41
	7.12	5.01	4.16	3.68	3.37	3.15	2.98	2.85	2.75	2.66	2.59	2.53	2.43	2.35	2.23	2.15	2.06	1.96	1.90	1.82	1.78	1.71	1.66	1.64
60	4.00	3.15	2.76	2.52	2.37	2.25	2.17	2.10	2.04	1.99	1.95	1.92	1.86	1.81	1.75	1.70	1.65	1.59	1.56	1.50	1.48	1.44	1.41	1.39
	7.08	4.98	4.13	3.65	3.34	3.12	2.95	2.82	2.72	2.63	2.56	2.50	2.40	2.32	2.20	2.12	2.03	1.93	1.87	1.79	1.74	1.68	1.63	1.60
65	3.99	3.14	2.75	2.51	2.36	2.24	2.15	2.08	2.02	1.98	1.94	1.90	1.85	1.80	1.73	1.68	1.63	1.57	1.54	1.49	1.46	1.42	1.39	1.37
	7.04	4.95	4.10	3.62	3.31	3.09	2.93	2.79	2.70	2.61	2.54	2.47	2.37	2.30	2.18	2.09	2.00	1.90	1.84	1.76	1.71	1.64	1.60	1.56
70	3.98	3.13	2.74	2.50	2.35	2.23	2.14	2.07	2.01	1.97	1.93	1.89	1.84	1.79	1.72	1.67	1.62	1.56	1.53	1.47	1.45	1.40	1.37	1.35
	7.01	4.92	4.08	3.60	3.29	3.07	2.91	2.77	2.67	2.59	2.51	2.45	2.35	2.28	2.15	2.07	1.98	1.88	1.82	1.74	1.69	1.62	1.56	1.53
80	3.96	3.11	2.72	2.48	2.33	2.21	2.12	2.05	1.99	1.95	1.91	1.88	1.82	1.77	1.70	1.65	1.60	1.54	1.51	1.45	1.42	1.38	1.35	1.32
	6.96	4.88	4.04	3.56	3.25	3.04	2.87	2.74	2.64	2.55	2.48	2.41	2.32	2.24	2.11	2.03	1.94	1.84	1.78	1.70	1.65	1.57	1.52	1.49

0.05 (light row) and 0.01 (dark row) points for the distribution of F

Degrees of freedom for greater mean square (numerator)

Degrees of freedom for lesser mean square	1	2	3	4	5	6	7	8	9	10	11	12	14	16	20	24	30	40	50	75	100	200	500	∞
100	3.94 / 6.90	3.09 / 4.82	2.70 / 3.98	2.46 / 3.51	2.30 / 3.20	2.19 / 2.99	2.10 / 2.82	2.03 / 2.69	1.97 / 2.59	1.92 / 2.51	1.88 / 2.43	1.85 / 2.36	1.79 / 2.26	1.75 / 2.19	1.68 / 2.06	1.63 / 1.98	1.57 / 1.89	1.51 / 1.79	1.48 / 1.73	1.42 / 1.64	1.39 / 1.59	1.34 / 1.51	1.30 / 1.46	1.28 / 1.43
125	3.92 / 6.84	3.07 / 4.78	2.68 / 3.94	2.44 / 3.47	2.29 / 3.17	2.17 / 2.95	2.08 / 2.79	2.01 / 2.65	1.95 / 2.56	1.90 / 2.47	1.86 / 2.40	1.83 / 2.33	1.77 / 2.23	1.72 / 2.15	1.65 / 2.03	1.60 / 1.94	1.55 / 1.85	1.49 / 1.75	1.45 / 1.68	1.39 / 1.59	1.36 / 1.54	1.31 / 1.46	1.27 / 1.40	1.25 / 1.37
150	3.91 / 6.81	3.06 / 4.75	2.67 / 3.91	2.43 / 3.44	2.27 / 3.13	2.16 / 2.92	2.07 / 2.76	2.00 / 2.62	1.94 / 2.53	1.89 / 2.44	1.85 / 2.37	1.82 / 2.30	1.76 / 2.20	1.71 / 2.12	1.64 / 2.00	1.59 / 1.91	1.54 / 1.83	1.47 / 1.72	1.44 / 1.66	1.37 / 1.56	1.34 / 1.51	1.29 / 1.43	1.25 / 1.37	1.22 / 1.33
200	3.89 / 6.76	3.04 / 4.71	2.65 / 3.88	2.41 / 3.41	2.26 / 3.11	2.14 / 2.90	2.05 / 2.73	1.98 / 2.60	1.92 / 2.50	1.87 / 2.41	1.83 / 2.34	1.80 / 2.28	1.74 / 2.17	1.69 / 2.09	1.62 / 1.97	1.57 / 1.88	1.52 / 1.79	1.45 / 1.69	1.42 / 1.62	1.35 / 1.53	1.32 / 1.48	1.26 / 1.39	1.22 / 1.33	1.19 / 1.28
400	3.86 / 6.70	3.02 / 4.66	2.62 / 3.83	2.39 / 3.36	2.23 / 3.06	2.12 / 2.85	2.03 / 2.69	1.96 / 2.55	1.90 / 2.46	1.85 / 2.37	1.81 / 2.29	1.78 / 2.23	1.72 / 2.12	1.67 / 2.04	1.60 / 1.92	1.54 / 1.84	1.49 / 1.74	1.42 / 1.64	1.38 / 1.57	1.32 / 1.47	1.28 / 1.42	1.22 / 1.32	1.16 / 1.24	1.13 / 1.19
1000	3.85 / 6.66	3.00 / 4.62	2.61 / 3.80	2.38 / 3.34	2.22 / 3.04	2.10 / 2.82	2.02 / 2.66	1.95 / 2.53	1.89 / 2.43	1.84 / 2.34	1.80 / 2.26	1.76 / 2.20	1.70 / 2.09	1.65 / 2.01	1.58 / 1.89	1.53 / 1.81	1.47 / 1.71	1.41 / 1.61	1.36 / 1.54	1.30 / 1.44	1.26 / 1.38	1.19 / 1.28	1.13 / 1.19	1.08 / 1.11
∞	3.84 / 6.64	2.99 / 4.60	2.60 / 3.78	2.37 / 3.32	2.21 / 3.02	2.09 / 2.80	2.01 / 2.64	1.94 / 2.51	1.88 / 2.41	1.83 / 2.32	1.79 / 2.24	1.75 / 2.18	1.69 / 2.07	1.64 / 1.99	1.57 / 1.87	1.52 / 1.79	1.46 / 1.69	1.40 / 1.59	1.35 / 1.52	1.28 / 1.41	1.24 / 1.36	1.17 / 1.25	1.11 / 1.15	1.00 / 1.00

Reprinted by permission from *Statistical Methods*, Seventh Edition, by George W. Snedecor and William G. Cochran. Copyright © 1980 by The Iowa State University Press, Ames, Iowa 50010.

Glossary

Operators

$+$, Σ	Addition	(11, 12)
$-$	Subtraction	(11)
$\times$, $\cdot$, $()()$	Multiplication	(11)
$\div$, $/$,	Division	(9)
^{2}, $()^2$	Square	(9)
$\sqrt{}$	Square root	(9, 15)
$!$	Factorial	(13)

Relations

$=$	Equal to	(11)
$>$	Greater than	(11)
$\geq$	Greater than or equal to	(11)
$<$	Less than	(11)
$\leq$	Less than or equal to	(11)
$\neq$	Not equal to	(11)

Variables

X	A variable	(14)
Y	A different variable	(14)
z	A standardized score, z score	(90–93)

Constants

X_i	A given score on the variable X	(12–14)
Y_i	A given score on the variable Y	(12–14)
z_i	A given score on the standardized variable, expressed in standard deviation units	(92)
r	The number of rows in a given contingency table	(15, 175)
k	The number of columns in a given contingency table	(15, 175)
N	The population size	(85–86)
n	The sample size	(85–86)
df	Degrees of freedom	(154–155)
α	Alpha, a selected level of significance	(156–159)

General statistical results

f	Frequency	(36)
f_o	Observed frequency	(170–171)
f_e	Expected frequency	(170–171)
p	Proportion	(39–40)
p	Probability	(40)
%	Percentage	(40–41)

Central tendency

Mo	Mode	(70)
Md	Median	(71)

Page numbers in parentheses indicate location in the text.

| $\bar{Y}$ | Mean of a sample   (72) |
| μ | Mean of a population   (72) |

Variation

v	Variation rat.o (82–84)
range	Range (84–85)
σ^2	Variance of a population (85–86)
s^2	Variance of a sample, to estimate the population variance (85–86)
σ	Standard deviation of a population (88–89)
s	Standard deviation of a sample, to estimate the population standard deviation (88–89)

Association

λ	Lambda (122–125)
γ	Gamma (125–127)
r	Pearson's correlation coefficient of a sample (127–133)
ρ	Pearson's correlation coefficient for a population (127)
r^2	Pearson's correlation coefficient squared of a sample (127, 132–133)
ρ^2	Pearson's correlation coefficient squared of a population (127)

Hypothesis testing

H_0	Null hypothesis (170)
H_R	Research hypothesis (171)
χ^2	Chi-square, a random variable with a chi-square distribution (170–173)
t	A random variable with a t distribution (178–181)
z	A standard normal random variable with a normal distribution (181, 214)
F	A random variable with an F distribution (193–198)
TSS	Total sum of squares (194)
MS	Mean square (194–195)
Z	Fisher Z transformation (214–218)

Index